Small Divisor Problem in the Theory of Three-Dimensional Water Gravity Waves

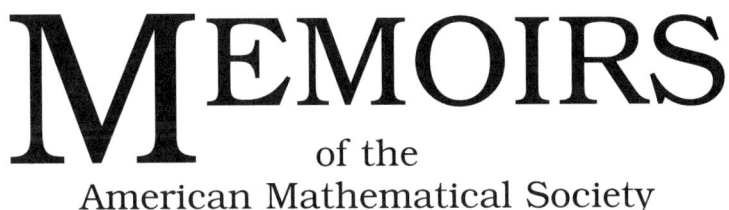

Memoirs
of the
American Mathematical Society

Number 940

Small Divisor Problem in the Theory of Three-Dimensional Water Gravity Waves

Gérard Iooss
Pavel I. Plotnikov

July 2009 • Volume 200 • Number 940 (fifth of 6 numbers) • ISSN 0065-9266

American Mathematical Society
Providence, Rhode Island

2000 *Mathematics Subject Classification.* Primary 76B15, 47J15, 35S15, 76B07.

Library of Congress Cataloging-in-Publication Data

Iooss, Gérard.
  Small divisor problem in the theory of three-dimensional water gravity waves / Gérard Iooss and Pavel I. Plotnikov.
      p. cm. — (Memoirs of the American Mathematical Society, ISSN 0065-9266 ; no. 940)
  "Volume 200, number 940 (fifth of 6 numbers)."
  Includes bibliographical references and index.
  ISBN 978-0-8218-4382-6 (alk. paper)
  1. Water waves. 2. Gravity waves. 3. Multiphase flow—Mathematical models. 4. Pseudo-differential operators. 5. Boundary value problems. 6. Bifurcation theory. 7. Small divisors. I. Plotnikov, Pavel I. II. Title.

QA922.L66 2009
532′.593—dc22
                                                                                            2009008894

Memoirs of the American Mathematical Society

This journal is devoted entirely to research in pure and applied mathematics.

Subscription information. The 2009 subscription begins with volume 197 and consists of six mailings, each containing one or more numbers. Subscription prices for 2009 are US$709 list, US$567 institutional member. A late charge of 10% of the subscription price will be imposed on orders received from nonmembers after January 1 of the subscription year. Subscribers outside the United States and India must pay a postage surcharge of US$65; subscribers in India must pay a postage surcharge of US$95. Expedited delivery to destinations in North America US$57; elsewhere US$160. Each number may be ordered separately; *please specify number* when ordering an individual number. For prices and titles of recently released numbers, see the New Publications sections of the *Notices of the American Mathematical Society*.

Back number information. For back issues see the *AMS Catalog of Publications*.

Subscriptions and orders should be addressed to the American Mathematical Society, P. O. Box 845904, Boston, MA 02284-5904, USA. *All orders must be accompanied by payment.* Other correspondence should be addressed to 201 Charles Street, Providence, RI 02904-2294, USA.

Copying and reprinting. Individual readers of this publication, and nonprofit libraries acting for them, are permitted to make fair use of the material, such as to copy a chapter for use in teaching or research. Permission is granted to quote brief passages from this publication in reviews, provided the customary acknowledgment of the source is given.

Republication, systematic copying, or multiple reproduction of any material in this publication is permitted only under license from the American Mathematical Society. Requests for such permission should be addressed to the Acquisitions Department, American Mathematical Society, 201 Charles Street, Providence, Rhode Island 02904-2294, USA. Requests can also be made by e-mail to reprint-permission@ams.org.

Memoirs of the American Mathematical Society (ISSN 0065-9266) is published bimonthly (each volume consisting usually of more than one number) by the American Mathematical Society at 201 Charles Street, Providence, RI 02904-2294, USA. Periodicals postage paid at Providence, RI. Postmaster: Send address changes to Memoirs, American Mathematical Society, 201 Charles Street, Providence, RI 02904-2294, USA.

© 2009 by the American Mathematical Society. All rights reserved.
Copyright of individual articles may revert to the public domain 28 years after publication. Contact the AMS for copyright status of individual articles.
This publication is indexed in *Science Citation Index*®, *SciSearch*®, *Research Alert*®, *CompuMath Citation Index*®, *Current Contents*®/*Physical, Chemical & Earth Sciences*.
Printed in the United States of America.

∞ The paper used in this book is acid-free and falls within the guidelines established to ensure permanence and durability.
Visit the AMS home page at http://www.ams.org/

10 9 8 7 6 5 4 3 2 1   14 13 12 11 10 09

Contents

Chapter 1. Introduction   1
   1.1. Presentation and History of the Problem   1
   1.2. Formulation of the Problem   4
   1.3. Results   7
   1.4. Mathematical background   9
   1.5. Structure of the paper   9

Chapter 2. Formal Solutions   15
   2.1. Differential of $\mathcal{G}_\eta$ 15
   2.2. Linearized equations at the origin and dispersion relation   15
   2.3. Formal computation of 3-dimensional waves   17
   2.4. Geometric pattern of diamond waves   20

Chapter 3. Linearized Operator   23
   3.1. Linearized system in $(\psi, \eta) \neq 0$ 23
   3.2. Pseudodifferential operators and diffeomorphism of the torus   25
   3.3. Main orders of the diffeomorphism and coefficient ν 33

Chapter 4. Small Divisors. Estimate of $\mathfrak{L}$– Resolvent   35
   4.1. Proof of Theorem 4.10   41

Chapter 5. Descent Method-Inversion of the Linearized Operator   51
   5.1. Descent method   52
   5.2. Proof of Theorem 5.1   61
   5.3. Verification of assumptions of Theorem 5.1   66
   5.4. Inversion of $\mathcal{L}$ 68

Chapter 6. Nonlinear Problem. Proof of Theorem 1.3   71

Appendix A. Analytical study of $\mathcal{G}_\eta$ 75
   A.1. Computation of the differential of $\mathcal{G}_\eta$ 75
   A.2. Second order Taylor expansion of $\mathcal{G}_\eta$ in $\eta = 0$ 77

Appendix B. Formal computation of 3-dimensional waves   79
   B.1. Formal Fredholm alternative   79
   B.2. Bifurcation equation   81

Appendix C. Proof of Lemma 3.6   87

Appendix D.	Proofs of Lemmas 3.7 and 3.8	89
Appendix E.	Distribution of Numbers $\{\omega_0 n^2\}$	93
Appendix F.	Pseudodifferential Operators	99
Appendix G.	Dirichlet-Neumann Operator	107
Appendix H.	Proof of Lemma 5.8	119
Appendix I.	Fluid particles dynamics	123
Bibliography		127

Abstract

We consider doubly-periodic travelling waves at the surface of an infinitely deep perfect fluid, only subjected to gravity g and resulting from the nonlinear interaction of two simply periodic travelling waves making an angle 2θ between them.

Denoting by $\mu = gL/c^2$ the dimensionless bifurcation parameter (L is the wave length along the direction of the travelling wave and c is the velocity of the wave), bifurcation occurs for $\mu = \cos\theta$. For non-resonant cases, we first give a large family of formal three-dimensional gravity travelling waves, in the form of an expansion in powers of the amplitudes of two basic travelling waves. "Diamond waves" are a particular case of such waves, when they are symmetric with respect to the direction of propagation.

The main object of the paper is the proof of existence of such symmetric waves having the above mentioned asymptotic expansion. Due to the *occurence of small divisors*, the main difficulty is the inversion of the linearized operator at a non trivial point, for applying the Nash Moser theorem. This operator is the sum of a second order differentiation along a certain direction, and an integro-differential operator of first order, both depending periodically of coordinates. It is shown that for almost all angles θ, the 3-dimensional travelling waves bifurcate for a set of "good" values of the bifurcation parameter having asymptotically a full measure near the bifurcation curve in the parameter plane (θ, μ).

Received by the editor November 10, 2005.

2000 *Mathematics Subject Classification.* 76B15; 47J15; 35S15; 76B07.

Key words and phrases. nonlinear water waves; small divisors; bifurcation theory; pseudodifferential operators; traveling gravity waves; short crested waves.

CHAPTER 1

Introduction

1.1. Presentation and History of the Problem

We consider small-amplitude three-dimensional doubly periodic travelling gravity waves on the free surface of a perfect fluid. These *unforced* waves appear in literature as steady 3-dimensional water waves, since they are steady in a suitable moving frame. The fluid layer is supposed to be infinitely deep, and the flow is irrotational only subjected to gravity. The bifurcation parameter is the horizontal phase velocity, the infinite depth case being not essentially different from the finite depth case, except for very degenerate situations that we do not consider here. The essential difficulty here, with respect to the existing literature is that *we assume the absence of surface tension*. Indeed the surface tension plays a major role in all existing proofs for three-dimensional travelling gravity-capillary waves, and when the surface tension is very small, which is the case in many usual situations, this implies a reduced domain of validity of these results.

In 1847 Stokes [**Sto**] gave a nonlinear theory of *two-dimensional* travelling gravity waves, computing the flow up to the cubic order of the amplitude of the waves, and the first mathematical proofs for such periodic two-dimensional waves are due to Nekrasov [**N**], Levi-Civita [**Le**] and Struik [**Str**] about 80 years ago. Mathematical progress on the study of *three-dimensional* doubly periodic water waves came much later. In particular, to our knowledge, first formal expansions in powers of the amplitude of three-dimensional travelling waves can be found in papers [**Fu**] and [**Sr**]. One can find many references and results of research on this subject in the review paper of Dias and Kharif [**DiK**] (see section 6). The work of Reeder and Shinbrot (1981)[**ReSh**] represents a big step forward. These authors consider symmetric diamond patterns, resulting from (horizontal) wave vectors belonging to a lattice Γ' (dual to the spatial lattice Γ of the doubly periodic pattern) spanned by two wave vectors K_1 and K_2 with the *same length*, the velocity of the wave being in the direction of the bissectrix of these two wave vectors, taken as the x_1 horizontal axis. We give in Figure 1 two examples of patterns for these waves (see the detailed comment about these pictures at the end of subsection 2.4). These waves also appear in the literature as "short crested waves" (see Roberts and Schwartz [**RoSc**], Bridges, Dias, Menasce [**BDM**] for an extensive discussion on various situations and numerical computations). If we denote by θ the angle between K_1 and the x_1-

axis, Reeder and Shinbrot proved that bifurcation to diamond waves occurs provided the angle θ is not too close to 0 or to $\pi/2$, and provided that the *surface tension is not too small*. In addition their result is only valid outside a "bad" set in the parameter space, corresponding to resonances, a quite small set indeed. This means that if one considers the dispersion relation $\Delta(K, \mathbf{c}) = 0$, where K and $\mathbf{c}$ are respectively a wave vector and the velocity of the travelling wave, then there is no resonance if for the critical value of the velocity $\mathbf{c}_0$ there are only the four solutions $\pm K_1, \pm K_2$ of the dispersion equation, for $K \in \Gamma'$ (i.e. for K being any integer linear combination of K_1 and K_2). The fact that the surface tension is supposed not to be too small is essential for being able to use Lyapunov-Schmidt technique, and the authors mention a *small divisor problem if there is no surface tension*, as computed for example in [**RoSc**]. Notice that the existence of spatially bi-periodic *gravity* water waves was proved by Plotnikov in [**OM**], [**P**] in the case of finite depth and for fixed rational values of $gL/c^2 \tan\theta$, where g, L, c are respectively the acceleration of gravity, the wave length in the direction of propagation, and the velocity of the wave. Indeed, such a special choice of parameters avoids resonances and the small divisor problem, because the pseudo-inverse of the linearized operator is bounded.

Craig and Nicholls (2000) [**CrN2**] used the hamiltonian formulation introduced by Zakharov [**Z**], in coupling the Lyapunov-Schmidt technique with a variational method on the bifurcation equation. Still in the presence of surface tension, they could suppress the restriction of Reeder and Shinbrot on the "bad" resonance set in parameter space, but they pay this complementary result in losing the smoothness of the solutions. Among other results, the other paper by Craig and Nicholls (2002) [**CrN1**] gives the principal parts of "simple" doubly periodic waves (i.e. in the non resonant cases), expanded in Taylor series, taking into account the two-dimensions of the parameter $\mathbf{c}$. They emphasize the fact that this expansion is only formal in the absence of surface tension.

Mathematical results of another type are obtained in using "spatial dynamics", in which one of the horizontal coordinates (the distinguished direction) plays the role of a time variable, as was initiated by Kirchgässner [**K**] and extensively applied to two-dimensional water wave problems (see a review in [**DiIo**]). The advantage of this method is that one does not choose the behavior of the solutions in the direction of the distinguished coordinate, and solutions periodic in this coordinate are a particular case, as well as quasi-periodic or localized solutions (solitary waves). In this framework one may a priori assume periodicity in a direction transverse to the distinguished direction, and a periodic solution in the distinguished direction is automatically doubly periodic. The first mathematical results obtained by this method, containing 3-dimensional doubly periodic travelling waves, start with Haragus, Kirchgässner, Groves and Mielke (2001) [**GM**], [**G**], [**HK**], generalized by Groves and Haragus (2003) [**GH**]. They use a hamiltonian formulation and center manifold reduction. This is essentially based

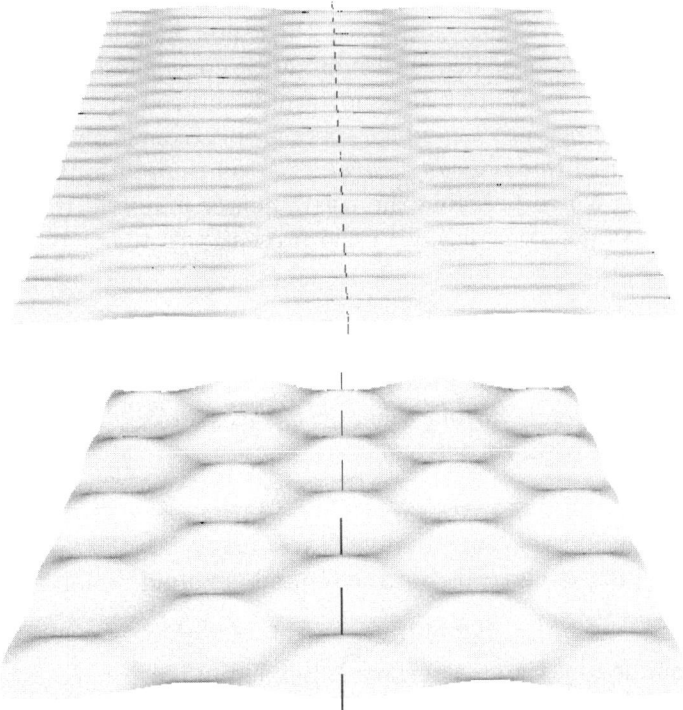

FIGURE 1. 3-dim travelling wave, the elevation $\eta_\varepsilon^{(2)}$ is computed with formula (1.10). Top: $\theta = 11.3°, \tau = 1/5, \varepsilon = 0.8\mu_c$; bottom: $\theta = 26.5°, \tau = 1/2, \varepsilon = 0.6\mu_c$. The dashed line is the direction of propagation of the waves. Crests are dark and troughs are grey.

on the fact that the spectrum of the linearized operator is discrete and has only a finite number of eigenvalues on the imaginary axis. These eigenvalues are related with the dispersion relation mentioned above. Here, one component (or multiples of such a component) of the wave vector K is imposed in a direction transverse to the distinguished one, and there is no restriction for the component of K in the distinguished direction, which, in solving the dispersion relation, gives the eigenvalues of the linearized operator on the imaginary axis. The resonant situations, in the terminology of Craig and Nicholls correspond here to more than one pair of eigenvalues on the imaginary axis, (in addition to the origin). In all cases it is known that the largest eigenvalue on the imaginary axis leads to a family of periodic solutions, via the Lyapunov center theorem (hamiltonian case), so, here again, there is no restriction on the resonant set in the parameter space at a fixed finite depth. The only restriction with this formulation is that it is necessary to assume that the depth of the fluid layer is finite. This ensures that the spectrum of the linearized operator has a spectral gap near the imaginary axis, which allows to use the center manifold reduction method. In fact if we restrict the study to periodic solutions as here, the center manifold reduction is not necessary, and the infinite depth case might be considered in using an extension of the proof of Lyapunov-Devaney center theorem in the spirit of [**I**], in this case where 0 belongs to the continuous spectrum. However, it appears that the *number of imaginary eigenvalues becomes infinite when the surface tension cancels*, which prevents the use of center manifold reduction in the limiting case we are considering in the present paper, not only because of the infinite depth.

1.2. Formulation of the Problem

Since we are looking for waves travelling with velocity **c**, let us consider *the system in the moving frame* where the waves look steady. Let us denote by φ the potential defined by

$$\varphi = \phi - \mathbf{c} \cdot X,$$

where ϕ is the usual velocity potential, $X = (x_1, x_2)$ is the 2-dim horizontal coordinate, x_3 is the vertical coordinate, and the fluid region is

$$\Omega = \{(X, x_3) : -\infty < x_3 < \eta(X)\},$$

which is bounded by the free surface Σ defined by

$$\Sigma = \{(X, x_3) : x_3 = \eta(X)\}.$$

We also make a scaling in choosing $|\mathbf{c}|$ for the velocity scale, and L for a length scale (to be chosen later), and we still denote by (X, x_3) the new coordinates, and by φ, η the unknown functions. Now defining the parameter $\mu = \frac{gL}{c^2}$ (the Froude number is $\frac{c}{\sqrt{gL}}$) where g denotes the acceleration of gravity, and **u** the unit vector in the direction of **c**, the system reads

(1.1) $$\Delta \varphi = 0 \text{ in } \Omega,$$

(1.2) $$\nabla_X \eta \cdot (\mathbf{u} + \nabla_X \varphi) - \frac{\partial \varphi}{\partial x_3} = 0 \text{ on } \Sigma,$$

(1.3) $$\mathbf{u} \cdot \nabla_X \varphi + \frac{(\nabla \varphi)^2}{2} + \mu \eta = 0 \text{ on } \Sigma,$$

$$\nabla \varphi \to 0 \text{ as } x_3 \to -\infty.$$

Hilbert spaces of periodic functions. We specialize our study to *spatially periodic 3-dimensional travelling waves*, i.e. the solutions η and φ are *bi-periodic in* X. This means that there are two independent wave vectors $K_1, K_2 \in \mathbb{R}^2$ generating a lattice

$$\Gamma' = \{K = n_1 K_1 + n_2 K_2 : n_j \in \mathbb{Z}\},$$

and a dual lattice Γ of periods in $\mathbb{R}^2$ such that

$$\Gamma = \{\lambda = m_1 \lambda_1 + m_2 \lambda_2 : m_j \in \mathbb{Z}, \lambda_j \cdot K_l = 2\pi \delta_{jl}\}.$$

The Fourier expansions of η and φ are in terms of $e^{iK \cdot X}$, where $K \in \Gamma'$ and $K \cdot \lambda = 2n\pi$, $n \in \mathbb{Z}$, for $\lambda \in \Gamma$. *The situation we consider in the* *further analysis, is with a lattice Γ' generated by the symmetric wave vectors $K_1 = (1, \tau)$, $K_2 = (1, -\tau)$. In such a case the functions on $\mathbb{R}^2/\Gamma$ are $2\pi-$ periodic in x_1, $2\pi/\tau$ periodic in x_2, and invariant under the shift $(x_1, x_2) \mapsto (x_1 + \pi, x_2 + \pi/\tau)$ (and conversely). We define the Fourier coefficients of a bi- periodic function u on such lattice Γ_τ by*

$$\widehat{u}(k) = \frac{\sqrt{\tau}}{2\pi} \int_{[0,2\pi]\times[0,2\pi/\tau]} u(X) \exp(-ik \cdot X) dX.$$

For $m \geq 0$ we denote by $H^m(\mathbb{R}^2/\Gamma)$ the Sobolev space of bi-periodic functions of $X \in \mathbb{R}^2/\Gamma$ which are square integrable on a period, with their partial derivatives up to order m, and we can choose the norm as

$$||u||_m = \left(\sum_{k \in \Gamma'} (1 + |k|)^{2m} |\widehat{u}(k)|^2 \right)^{1/2}.$$

Operator equations. Now, we reduce the above system for (φ, η) to a system of two scalar equations in choosing the new unknown function

$$\psi(X) = \varphi(X, \eta(X)),$$

and we define the Dirichlet-Neumann operator $\mathcal{G}_\eta$ by

$$(1.4) \quad \begin{aligned} \mathcal{G}_\eta \psi &= \sqrt{1 + (\nabla_X \eta)^2} \frac{d\varphi}{dn}\Big|_{x_3=\eta(X)} \\ &= \frac{\partial \varphi}{\partial x_3}\Big|_{x_3=\eta(X)} - \nabla_X \eta \cdot \nabla_X \varphi \end{aligned}$$

where n is normal to Σ, exterior to Ω, and φ is the solution of the $\eta-$ dependent Dirichlet problem

$$\begin{aligned} \Delta \varphi &= 0, \quad x_3 < \eta(X) \\ \varphi &= \psi, \quad x_3 = \eta(X), \\ \nabla \varphi &\to 0 \text{ as } x_3 \to -\infty. \end{aligned}$$

Notice that this definition of $\mathcal{G}_\eta$ follows [**La**] and insures the selfadjointness and positivity of this linear operator in $L^2(\mathbb{R}^2/\Gamma)$ (see Appendix A.1). Our definition differs from another usual way of defining the Dirichlet - Neumann operator without the square root in factor in (1.4). Now we have the identity(1.4) and the system to solve reads

$$(1.5) \quad \mathcal{F}(U, \mu, \mathbf{u}) = 0, \quad \mathcal{F} = (\mathcal{F}_1, \mathcal{F}_2),$$

where $U = (\psi, \eta)$, and

$$(1.6) \quad \mathcal{F}_1(U, \mu, \mathbf{u}) = : \mathcal{G}_\eta(\psi) - \mathbf{u} \cdot \nabla_X \eta,$$

$$(1.7) \quad \mathcal{F}_2(U, \mu, \mathbf{u}) = : \mathbf{u} \cdot \nabla_X \psi + \mu \eta + \frac{(\nabla \psi)^2}{2} + $$
$$- \frac{1}{2(1 + (\nabla_X \eta)^2)} \{\nabla_X \eta \cdot (\nabla_X \psi + \mathbf{u})\}^2.$$

Let us define the 2-components function space
$$\mathbb{H}^m(\mathbb{R}^2/\Gamma) = H_0^m(\mathbb{R}^2/\Gamma) \times H^m(\mathbb{R}^2/\Gamma)$$
We denote the norm of U in $\mathbb{H}^m(\mathbb{R}^2/\Gamma)$ by
$$||U||_m = ||\psi||_{H^m} + ||\eta||_{H^m},$$
where H_0^m means functions with 0 average, and $U = (\psi, \eta)$. The 0 average condition comes from the fact that the value ψ of the potential is defined up to an additive constant (easily checked in equations (1.6), (1.7)). Moreover, the average of the right hand side of (1.6) is 0 as it can be easily checked (this is proved for instance in [**CrN1**]). We have the following

LEMMA 1.1. *For any fixed $m \geq 3$, the mapping*
$$(U, \mu, \mathbf{u}) \mapsto \mathcal{F}(U, \mu, \mathbf{u}) \ \ is \ \ C^\infty : \mathbb{H}^m(\mathbb{R}^2/\Gamma) \times \mathbb{R} \times \mathbb{S}_1 \to \mathbb{H}^{m-1}(\mathbb{R}^2/\Gamma)$$
in the neighborhood of $\{0\} \times \mathbb{R} \times \mathbb{S}_1$. Moreover $\mathcal{F}(\cdot, \mu, \mathbf{u})$ is equivariant under translations of the plane:
$$\mathcal{T}_\mathbf{v} \mathcal{F}(U, \mu, \mathbf{u}) = \mathcal{F}(\mathcal{T}_\mathbf{v} U, \mu, \mathbf{u})$$
where
$$\mathcal{T}_\mathbf{v} U(X) = U(X + \mathbf{v}).$$
In addition, there is $M_3 > 0$, such that for $||U||_3 \leq M_3$ and $|\mu| \leq M_3$, $\mathcal{F}$ satisfies for any $m \geq 3$ the "tame" estimate

(1.8) $$||\mathcal{F}(U, \mu, \mathbf{u})||_{m-1} \leq c_m(M_3)||U||_m,$$
where c_m only depends on m and M_3.

PROOF. The C^∞ smoothness of $(\psi, \eta) \mapsto \mathcal{G}_\eta(\psi) : \mathbb{H}^m(\mathbb{R}^2/\Gamma) \to H^{m-1}(\mathbb{R}^2/\Gamma)$ comes from the study of the Dirichlet-Neumann operator, see (A.1,A.2), and the properties of elliptic operators. This result is proved in particular by Craig and Nicholls in [**CrN2**], and by D.Lannes in [**La**]. Notice that $H^s(\mathbb{R}^2/\Gamma)$ is an algebra for $s > 1$. Notice that it is proved by Craig et al [**CrSS**] that the mapping $(\psi, \eta) \mapsto \mathcal{G}_\eta(\psi) : H^m(\mathbb{R}^2/\Gamma) \times C^m(\mathbb{R}^2/\Gamma) \to H^{m-1}(\mathbb{R}^2/\Gamma)$ is analytic and the authors give the explicit Taylor expansion near 0, with the same type of "tame" estimates that we shall use in the following chapters. We choose here to stay with $(\psi, \eta) \in \mathbb{H}^m(\mathbb{R}^2/\Gamma)$ and we just use the C^∞ smoothness of the mapping, in addition to the tame estimates (see [**La**]).

The equivariance of $\mathcal{F}$ under translations of the plane is obvious.

We refer to [**La**] for the proof of the following "tame" estimate, valid for any $k \geq 1$ (here simpler than in [**La**] since we have periodic functions and since there is no bottom wall)

(1.9) $$||\mathcal{G}_\eta(\psi)||_k \leq c_k(||\eta||_3)\{||\eta||_{k+1}||\psi||_3 + ||\psi||_{k+1}\},$$

necessary to get estimate (1.8). □

1.3. Results

We are now in a position to formulate the main result of this paper on the existence of non-linear diamond waves satisfying operator equation (1.5). We find an explicit solution to (1.5) in the vicinity of an approximate solution $U_\varepsilon^{(N)}$ which existence is stated in the following lemma restricted to "diamond waves", i.e to solutions belonging to the important subspace (still with Γ' generated by $(1, \pm\tau)$)

$$\mathbb{H}_{(S)}^k = \{U = (\psi, \eta) \in \mathbb{H}^k(\mathbb{R}^2/\Gamma) : \psi \text{ odd in } x_1, \text{ even in } x_2, \eta \text{ even in } x_1 \text{ and in } x_2\}.$$

For these solutions the unit vector $\mathbf{u}_0 = (1, 0)$ is fixed (see a more general statement at Theorem 2.3, with non necessarily symmetric formal solutions).

LEMMA 1.2. *Let $N \geq 3$ be an arbitrary positive number and the critical value of parameter $\mu_c(\tau) = (1+\tau^2)^{-1/2}$ is such that the dispersion equation $n^2 + \tau^2 m^2 = \mu_c^{-2} n^4$ has only the solution $(n, m) = (1, 1)$ in the circle $m^2 + n^2 \leq N^2$. Then approximate 3-dimensional diamond waves are given by*

$$(1.10) \qquad U_\varepsilon^{(N)} = (\psi, \eta)_\varepsilon^{(N)} = \sum_{1 \leq p \leq N} \varepsilon^p U^{(p)} \in \mathbb{H}_{(S)}^k, \text{ for any } k,$$

$$U^{(1)} = (\sin x_1 \cos \tau x_2, \frac{-1}{\mu_c} \cos x_1 \cos \tau x_2), \quad \mu_\varepsilon^{(N)} = \mu_c + \tilde{\mu}, \quad \tilde{\mu} = \mu_1 \varepsilon^2 + O(\varepsilon^4),$$

where

$$\mu_1 = \left(\frac{1}{4\mu_c^3} - \frac{1}{2\mu_c^2} - \frac{3}{4\mu_c} + 2 + \frac{\mu_c}{2} - \frac{9}{4(2-\mu_c)}\right),$$

and where for any k,

$$\mathcal{F}(U_\varepsilon^{(N)}, \mu_\varepsilon^{(N)}, \mathbf{u}_0) = \varepsilon^{N+1} Q_\varepsilon,$$

Q_ε *uniformly bounded in $\mathbb{H}_{(S)}^k$, with respect to ε. There is one critical value τ_c of τ such that $\mu_1(\tau_c) = 0$, and $\mu_1(\tau) < 0$ for $\tau < \tau_c$, $\mu_1(\tau) > 0$ for $\tau > \tau_c$.*

PROOF. The lemma is a particular case of the general Theorem 2.3 in the symmetric case. □

The following theorem on existence of $3D$-diamond waves is the main result of the paper (notice that $\tau = \tan\theta$)

THEOREM 1.3. *Let us choose arbitrary integers $l \geq 23$, $N \geq 3$ and a real number $\delta < 1$. Assume that*

$$\tau \in (\delta, 1/\delta), \quad \mu_c = (1+\tau^2)^{-1/2}.$$

Then there is a set $\mathfrak{N}$ of full measure in (0,1) with the following property. If $\mu_c \in \mathfrak{N}$ and $\tau \neq \tau_c$, then there exists a positive $\varepsilon_0 = \varepsilon_0(\mu_c, N, l, \delta)$ and a set $\mathcal{E} = \mathcal{E}(\mu_c, N, l, \delta)$ so that

$$\lim_{\varepsilon \to 0} \frac{2}{\varepsilon^2} \int_{\mathcal{E} \cap (0, \varepsilon)} s\, ds = 1,$$

and for every $\mu = \mu_\varepsilon^{(N)}$ with $\varepsilon \in \mathcal{E}$, equation (1.5) has a "diamond wave" type solution $U = U_\varepsilon^{(N)} + \varepsilon^N W_\varepsilon$ with $W_\varepsilon \in \mathbb{H}^l_{(S)}$. Moreover, $W : \mathcal{E} \to \mathbb{H}^l_{(S)}$ is a Lipschitz function cancelling at $\varepsilon = 0$, and for $\tau < \tau_c$ (resp. $\tau > \tau_c$), and when ε varies in $\mathcal{E}$, the parameter $\mu = \mu_\varepsilon^{(N)}$ runs over a measurable set of the interval $(\mu_{\varepsilon_0}^{(N)}, \mu_c)$ (resp. $(\mu_c, \mu_{\varepsilon_0}^{(N)})$) of asymptotically full measure near μ_c.

We can roughly express our result in considering the two-dimensional parameter plane (τ, μ) where $\tau = \tan \theta$, 2θ being the angle between the two basic wave vectors of same length generating the two-dimensional lattice Γ' dual of the lattice Γ of periods for the waves. The critical value $\mu_c(\tau)$ of μ $(= gL/c^2)$, where $\mu_c(\tau) = (1+\tau^2)^{-1/2} = \cos \theta$, corresponds to the solutions of the dispersion relation we consider here (in particular 3-dimensional diamond waves propagating in the direction of the bisectrix of the wave vectors). We show that for $\tau < \tau_c$ (≈ 2.48) the bifurcating (diamond) waves of size $O(|\mu - \mu_c(\tau)|^{1/2})$ occur for $\mu < \mu_c(\tau)$, while for $\tau > \tau_c$ it occurs for $\mu > \mu_c(\tau)$. We prove that bifurcation of these 3-dimensional waves occurs on half lines $\tau = const$ of the plane, with their origin on the critical curve, for "good" values of τ (which appear to be nearly all values of τ). Moreover, we prove that on each half line, these waves exist for "good" values of μ, this set of "good" values being asymptotically of full measure at the bifurcation point $\mu = \mu_c(\tau)$ (see Figure 2).

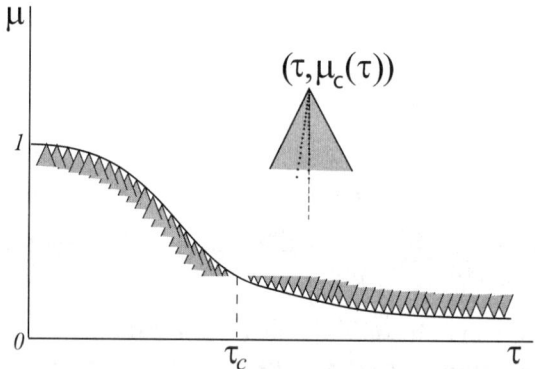

FIGURE 2. Small sectors where 3-dimensional waves bifurcate. Their vertices lie on the critical curve $\mu = \mu_c(\tau)$. The good set of points is asymptotically of full measure at the vertex on each half line (see the detail above). In the paper we only give the proof for each half line $\tau = const$ (dashed line on the figure)

Another way to describe our result is in terms of a bifurcation from a non isolated eigenvalue in the spectrum of the linearized operator at the origin. Indeed, for our critical values $(\tau, \mu_c(\tau))$ of the parameter, the differential at the origin is a selfadjoint operator with in general a non isolated 0 eigenvalue (see Theorem 4.1). Our result means that from each point $(\tau, \mu_c(\tau))$ where τ is chosen in a full measure set of $(0, \infty)$, a branch of solutions bifurcates in the following sense. Every half line $\tau = const$ with origin at the point $(\tau, \mu_c(\tau))$ and on the good side of this curve, contains a measurable set of points where the bi-periodic gravity waves exist, with an amplitude $O(|\mu - \mu_c|^{1/2})$, this set being asymptotically of full measure near $\mu_c(\tau)$.

In fact, we can improve our result in replacing the half lines mentioned above, by *small sectors centered on these half lines*. Each half line in each sector, with origin at the vertex of the sector, contains a measurable set where the bi-periodic gravity waves exist, with an amplitude $O(|\mu - \mu_c|^{1/2})$, this set being asymptotically of full measure near $\mu_c(\tau)$. The proof of such a result introduces many technicalities, which are not essential for the understanding of the paper. This complication is mainly due to the fact that we then need to work with a lattice Γ now depending on ε. We just mention in various places what is really needed for such an extension of the result proved here.

1.4. Mathematical background

There are some aspects of our method which deserve brief mention. First we use the Nash-Moser method, which is now an integral part of nonlinear analysis [**D**], for proving Theorem 1.3. The crucial point for the Nash-Moser method is to obtain a priori bounds on an approximate right-inverse of the partial derivative $\partial_U \mathcal{F}(U, \mu, \mathbf{u})$. As it is shown in Chapter 3, this problem is equivalent to the problem of invertibility of a second-order selfadjoint pseudodifferential operator with multiple characteristics. Second, we use the Moser theory of foliation on a torus [**M**] and the *invariant parametric representation* of the Dirichlet-Neumann operator to reduce the linearized equation to a canonical form with constant coefficients in the principal part. We employ a modification of the Weil Theorem [**W**] on uniform distribution of numbers $\{\omega n^2\}$ modulo 1 to deduce the effective estimates of *small divisors*, and hence to prove the invertibility of the principal part of the linearized operator. The most essential ingredient of our approach is the algebraic *descent method* [**IPT**], [**PT**] which allows to reduce the canonical pseudodifferential equation on 2-dimensional torus to a Fredholm-type equation.

1.5. Structure of the paper

Now we can explain the organization of the paper. In Chapter 2, we prove Theorem 2.3 which establishes the existence of approximate solutions under the form of power series of the amplitudes of the two incident monoperiodic travelling waves, corresponding to symmetric basic wave vectors. *The parameter is two-dimensional here*, due to the freedom in the direction of propagation of the three-dimensional wave. To show this result, we use a formal Lyapunov-Schmidt technique, assuming that the angle 2θ between the two basic wave vectors satisfies that $\tau = \tan\theta$ is such that the equation for positive integers (n, m)

$$n^2 + \tau^2 m^2 = n^4(1 + \tau^2)$$

has the unique solution $(n, m) = (1, 1)$ (non resonance property). In playing with scales and parameters, this condition is not restrictive among non resonant situations, which indeed represent the general case. In such a case, the kernel of the linearized operator at rest state (taken as the origin) is four-dimensional, and in using extensively the symmetries of the system (1.5), we obtain, for a fixed value of the bifurcation parameter $(\mu, \mathbf{u})$, doubly-periodic formal travelling gravity waves propagating in the direction $\mathbf{u}$. Limiting cases are the mono-periodic travelling waves corresponding to one of the basic wave vectors. The Lemma 1.2 is a particular case of the above theorem.

From now on, we restrict the study to solutions called "diamond waves", which are *symmetric with respect to the direction of propagation*, here the x_1- axis. In Chapter 3 we consider the linear operator $\mathcal{L}(U, \mu)$ corresponding to the differential of (1.5) at a non zero point in $\mathbb{H}^k_{(S)}$, which we need to invert for using the Nash-Moser theorem. The principal part of this operator is the symmetric sum

$$-\mathcal{J}^*(\frac{1}{\mathfrak{a}}\mathcal{J}\cdot) + \mathcal{G}_\eta, \quad \mathcal{J} = V \cdot \nabla,$$

of a second order derivative in the direction of a periodic vector field $V(X)$, and of the Dirichlet-Neumann operator which is integro-differential of first order, both parts depending periodically on coordinates. More precisely, $V = \mathbb{G}^{-1}(X)(\mathbf{u}_0 + \nabla\psi(X))$, where $\mathbb{G}(X)\,dX \cdot dX$ is the first fundamental form of the free surface. Recall that $\mathbb{G}$ is a covariant tensor field on Σ, and for the standard parametrization $x_3 = \eta(X)$, it is given by $\mathbb{G}(X) = 1 + \nabla\eta \otimes \nabla\eta$. It follows from the kinematic condition (1.2) that integral curves of the vector field $V(X)$ coincide with trajectories of liquid particles moving along Σ and submitted to the vertical gravity μ.

Chapter 3 is concerned with the first step of the long way towards the inversion of $\mathcal{L}$, which consists in finding a diffeomorphism of the torus for which the highest order terms of the operator $\mathcal{L}$ become constants (depending on the linearization point). We begin (Lemma 3.6) with the construction of a diffeomorphism which takes integral curves of the vector field V onto straight lines parallel to the abscissa axis. Being endowed with the Jacobi metric $ds^2 = (1/2 - \mu\eta(X))\mathbb{G}(X)\,dX \cdot dX$ the free surface becomes a Riemannian manifold on which the integral curves of V coincide with geodesics (see Appendix I). Hence, by Lemma 3.6, they form a geodesic foliation on Σ. Moreover, since the distance between each of these curves and the abscissa axis is finite, the foliation has a zero rotation number. It is at this point where the restriction to symmetric solutions (diamond waves) is necessary, since we don't know yet how to manage such a diffeomorphism in the non symmetric case, see [**M**] for discussion. Recall that the Moser Theorem [**M**] guarantees the existence of at least one geodesic for any given rotation number.

The second result of Chapter 3 is Theorem 3.5 which gives the parametric representation of the Dirichlet-Neumann operator in arbitrary coordinates Y on Σ so that a mapping $X = X(Y)$ is a diffeomorphism of a torus. It follows from this theorem that for any smooth periodic function $u(Y)$ and $\breve{u}(X) = u(Y(X))$, the Dirichlet-Neumann operator has the decomposition

$$\mathcal{G}_\eta \breve{u} = \breve{\mathcal{G}_1} u + \breve{\mathcal{G}_0} u + \breve{\mathcal{G}_{-1}} u,$$

in which $\mathcal{G}_i$ are pseudodifferential operators of order i. We give an explicit expression for their symbols in terms of the first fundamental form and the principal curvatures of the free surface. In particular, we show that, up to a positive invariant multiplier, the symbol of the operator $\mathcal{G}_1$ is equal to $\sqrt{\mathbb{G}(Y)^{-1} k \cdot k}$, and the real part of the symbol of $\mathcal{G}_0$ coincides with the difference between the sum of the principal curvatures and the normal curvature of Σ in the direction of $\mathbb{G}^{-1} k$. This leads to the interesting conclusion: *the manifold Σ is defined by its Dirichlet-Neumann operator up to translation and rotation of the embedding space.*

Combining Lemma 3.6 and Theorem 3.5 gives the main result of Chapter 3 – Theorem 3.4. This theorem ensures the existence of a diffeomorphism $X = X(Y)$ of the 2-dimensional torus, which brings the linearized operator to the canonical form

$$\mathfrak{L} + \mathfrak{H} = \mathfrak{L} + \mathfrak{A} \partial_{y_1} + \mathfrak{B} + \mathfrak{L}_{-1},$$

where the remainder $\mathfrak{L}_{-1}$ is of order -1, $\mathfrak{A}$ and $\mathfrak{B}$ are zero-order pseudodifferential operators, and the principal part

$$\mathfrak{L} = \nu \partial_{y_1}^2 + (-\Delta)^{1/2}, \quad \Delta = \partial_{y_1}^2 + \tau^2 \partial_{y_2}^2$$

is a selfadjoint pseudodifferential operator. Here the parameter ν depends on the point of linearization, with $\nu(0) = \nu_0 = \mu_c(\tau)^{-1}$ (Lemma 3.7).

In Chapter 4 we study the operator $\mathfrak{L}$ in many details, and give estimates on its resolvent in Sobolev spaces of bi-periodic functions which are odd in y_1, and even in y_2. We begin with the observation that for $\nu_0 = \mu_c(\tau)^{-1}$ and almost every positive τ, zero is a simple eigenvalue of the operator $\mathfrak{L}_0 = \nu_0 \partial_{y_1}^2 + (-\Delta)^{1/2}$ and

$$\|\mathfrak{L}_0^{-1} u\|_s \leq c(\tau, \alpha) \|u\|_{s+(1+\alpha)/2}$$

for all u orthogonal to the kernel of $\mathfrak{L}_0$ and $\alpha > 0$. Next we study the perturbation of its resolvent assuming that $\nu = \nu_0 - \varepsilon^2 \nu_1 + O(\varepsilon^3)$ and with a spectral parameter $\varkappa = O(\varepsilon^2)$, both being Lipschitz functions of a small parameter ε. Here we have a *small divisor* problem, and we meet the necessity to restrict the parameter values to "good ones", for being able to find suitable estimates. Calculations (Lemma 4.5) show that the resolvent of $\mathfrak{L}$ satisfies the estimate

(1.11) $$\|(\mathfrak{L} - \varkappa)^{-1}u\|_s \leq c\|u\|_{s+1}, \quad u \in (\ker \mathfrak{L}_0)^\perp,$$

if parameters ν and $\varkappa$ satisfy the quadratic Diophantine inequalities

(1.12) $$|\omega n^2 - m - C| \geq cn^{-2} \text{ for all positive integers } n, m,$$

where $\omega = \nu\tau^{-1}$ and $C = (2\nu\tau)^{-1} - \varkappa\tau^{-1}$. Note that there is a difference between linear and polynomial Diophantine approximations: in classic theory of linear Diophantine forms, see [**Ca**] for general theory, the integers for which "small divisors" are really small, form a sparse set in the integral lattice. This property was used in pioneering works of Siegel [**Si**] and in the Arnold proof of the Kolmogorov Theorem [**A**]. In contrast to the linear case, the couples (m, n), for which the left hand side of inequality (1.12) is small, can form clusters in $\mathbb{Z}^2$, and the problem of obtaining small divisors estimates becomes more complicated. It turns out that the validity of inequalities (1.12) with a constant c independent of the small parameter is a consequence of estimate

(1.13)
$$N^{-1} \text{ card } \{n : \omega_0 n^2 \text{ modulo } 1 \leq \varepsilon \text{ and } 1 \leq n \leq N\} \leq c\varepsilon \text{ for all } N \geq \varepsilon^{-\lambda}.$$

Recall that, by the Weil Theorem [**W**], [**Ca**], for each fixed ε, the left hand side tends to ε as $N \to \infty$. Hence the inequality holds true for some c depending on ε. In Appendix E we make this result more precise and prove the existence of absolute constant c such that inequality (1.13) is fulfilled for all $\lambda \geq 78$ and all intervals of length ε. This leads to the main result of this chapter – Theorem 4.2, which shows that with a suitable choice of the parameters, the resolvent operator provides a loss of one in the degree of differentiability. Moreover, estimate (1.11) holds true for all ε^2 in an asymptotically full measure set on every half line $\theta = const$ of the parameter plane, the origin of which being chosen arbitrarily in a full measure set, on the bifurcation curve $\mu = \cos\theta$.

In Chapter 5 we take into account all remaining terms of the linear operator $\mathfrak{L} + \mathfrak{H}$ and prove its invertibility with a loss of differentiability. The main difficulty is that the operator $\mathfrak{L} + \mathfrak{H}$ involves the principal part $\mathfrak{L}$, which inverse is unbounded, and arbitrary operators $\mathfrak{A}$, $\mathfrak{B}$ with "variable coefficients".

Most, if not all, existing results related to such problems were obtained by use of *the Fröhlich-Spencer method* proposed in [**FS**], cf [**OM, PY**], and developed by Craig and Wayne [**CrWa, Cr**] and Bourgain [**B1, B2**]. The basic idea of the method is a representation of operators in the form of infinite matrices with elements labelled by some lattice and block decompositions of this lattice. Let us use the operator $\mathfrak{L} + \mathfrak{H}$ to illustrate the main features of this method. First we have to replace a periodic function u by the sequence of its Fourier coefficients $\{\widehat{u}(k)\}$, $k \in \Gamma'$, and the operator $\mathfrak{L}$ by the diagonal matrix with the elements $L(k)$. Then we have to split the lattice Γ' into a "regular" part which consists of all k with "large" $L(k)$, and an "irregular" part which includes all k corresponding to "small" values of $L(k)$. Using the contraction mapping principle we can eliminate the "regular" component and reduce the inversion of $\mathfrak{L} + \mathfrak{H}$ to the inversion of an infinite matrix on the "irregular" subspace. The existence of an inverse to this matrix is established by using a special iteration process which is the core of the method. Note that the Fröhlich-Spencer method is working in our case only if $\mathfrak{A} = 0$.

Our approach is based on *the descent method* which was proposed in [**PT, IPT**] and dates back to the classic Floquet-Lyapunov theory. The descent method is a pure algebraic procedure which brings the canonical operator to an operator with constant coefficients and does not depend on the structure and spectral properties of the principal part $\mathfrak{L}$. The heart of the method is the following identity (Theorem 5.2)

$$(\mathfrak{L} + \mathfrak{A}\mathfrak{D}_1 + \mathfrak{B})(1 + \mathfrak{C})u = (1 + \mathfrak{E})(\mathfrak{L} - \varkappa)u + \mathfrak{F}u,$$

which holds true for all functions $u \in H^2(\mathbb{R}^2/\Gamma)$ odd in y_1. Here $\mathfrak{C}$ and $\mathfrak{E}$ are bounded operators in the Sobolev spaces of periodic functions $H^s(\mathbb{R}^2/\Gamma)$; the remainder $\mathfrak{F}$ is a bounded operator : $H^{s-1}(\mathbb{R}^2/\Gamma) \mapsto H^s(\mathbb{R}^2/\Gamma)$; the Floquet exponent $\varkappa$ has an explicit expression in terms of operators $\mathfrak{A}$ and $\mathfrak{B}$. Moreover, if $\|\mathfrak{A}\|, \|\mathfrak{B}\| \sim \varepsilon$, then $\|\mathfrak{C}\|, \|\mathfrak{E}\|, \|\mathfrak{F}\| \sim \varepsilon$ and $\varkappa = O(\varepsilon^2)$. The proof of these results constitutes Section 5.1 and Appendix F. The technique used is more general than the one used in [**IPT**], since we use here general properties of pseudodifferential operators, however taking into account of the symmetry properties of $\mathcal{L}$.

The descent method of algebraic character presented here, might be easily used for example on the one-dimensional KDV and Schrödinger equations, avoiding the heavy technicalities of the Fröhlich-Spencer method.

Thus we reduce the problem of the inversion of the canonical operator $\mathfrak{L} + \mathfrak{H}$ to the problem of the inversion of operator $\mathfrak{L} - \varkappa + \tilde{\mathfrak{L}}_{-1}$, where $\tilde{\mathfrak{L}}_{-1}$ is a smoothing remainder. It is then possible to use the result of Chapter 4 for inverting the full operator and to prove Theorem 5.1 – the main result on the existence and estimates of $(\mathfrak{L}+\mathfrak{H})^{-1}$. In particular, this theorem implies that if $\mathfrak{A}$ and $\mathfrak{B}$ are Lipschitz operator-valued functions of a small parameter ε, which vanish for $\varepsilon = 0$ and satisfy symmetry and metric conditions (Chapter 5), and if $\mathfrak{L} - \varkappa$ meets all requirements of Theorem 4.2, then for all ε^2 taken in an asymptotically full measure set, the resolvent has the representation

$$(\mathfrak{L} + \mathfrak{H})^{-1}(\varepsilon) = \frac{1}{\mathfrak{c}}\mathfrak{H}_0(\varepsilon) + \mathfrak{H}_1(\varepsilon),$$

in which operators $\mathfrak{H}_1(\varepsilon) : H^s(\mathbb{R}^2/\Gamma) \mapsto H^{s-1}(\mathbb{R}^2/\Gamma)$ are uniformly bounded in ε, and $\mathfrak{H}_0(\varepsilon)$ are bounded operators of rank 1, the coefficient $\mathfrak{c} = \varepsilon^2 @ + O(\varepsilon^3)$ being given by (5.9). We show at the end of the chapter (see Theorem 5.9), that the results apply to the linear operator $\mathcal{L}(U, \mu)$ corresponding to the differential of (1.5) at a non zero point in $\mathbb{H}^k_{(S)}$. In particular, we give the sufficient conditions which provide the existence of the bounded inverse $\mathcal{L}(U,\mu)^{-1} : H^{s+3}(\mathbb{R}/\Gamma) \mapsto H^s(\mathbb{R}/\Gamma)$.

Chapter 6 applies extensively the result proved in [**IPT**] concerning the Nash-Moser theorem with parameters in a Cantor set. The main result, which is the main result of the paper is Theorem 1.3 establishing the existence of smooth bi-periodic travelling gravity waves symmetric with respect to the direction of propagation, in the region of the parameter space mentioned above. Notice that, apart from the last chapter, which heavily rests upon the self contained Appendix N of [**IPT**], the rest of the paper is self contained, with some details of computations and basics on pseudodifferential operators put in the Appendix, for providing an easy reading.

CHAPTER 2

Formal Solutions

2.1. Differential of $\mathcal{G}_\eta$

In this section we study the structure of the operator $\mathcal{G}_\eta$, and we give useful formulas and estimates.

The following regularity property holds

LEMMA 2.1. *The differential $h \mapsto \partial_\eta \mathcal{G}_\eta[h]$ of G_η satisfies for η, ψ, h smooth enough bi-periodic functions*

(2.1) $\quad \partial_\eta \mathcal{G}_\eta[h]\psi = -\mathcal{G}_\eta(h\zeta) + \nabla_X \cdot \{(\zeta \nabla_X \eta - \nabla_X \psi)h\},$

(2.2) $\quad \zeta = \dfrac{1}{1+(\nabla_X \eta)^2}\{\mathcal{G}_\eta \psi + \nabla_X \eta \cdot \nabla_X \psi\}.$

Moreover, despite the apparent loss of derivatives for (ψ, η) in (2.1,2.2) we have for $\|U\|_3 \leq M_3$, the following tame estimate

$$\|\partial_\eta \mathcal{G}_\eta[h]\psi\|_k \leq c_k(M_3)\{\|h\|_{k+1} + \|U\|_{k+1}\|h\|_2\}.$$

PROOF. We refer to Appendix A.1 for the formula (2.1,2.2) already proved for instance in [**La**], and we also refer to [**La**] for the tame estimate. □

From the formulas of the above Lemma 2.1, we are now able to compute successive derivatives of $\mathcal{F}$. Observe that in (2.1,2.2) there is a loss of two derivatives for (ψ, η). In fact there is a compensation cancelling the dependence into the second order derivatives and we have the following Lemma which completes Lemma 1.1:

LEMMA 2.2. *For $\|U\|_3 \leq M_3$, and $|\mu| \leq M_3$ the following tame estimates hold (and analogous ones for higher order derivatives)*

$$\|\partial_U \mathcal{F}(U,\mu,\mathbf{u})[\delta U]\|_k \leq c_k(M_3)\{\|\delta U\|_{k+1} + \|U\|_{k+1}\|\delta U\|_3\},$$

$$\|\partial^2_{UU}\mathcal{F}(U,\mu,\mathbf{u})[\delta U_1, \delta U_2]\|_k \leq c_k(M_3)\{\|\delta U_1\|_{k+1}\|\delta U_2\|_3 +$$
$$+\|\delta U_2\|_{k+1}\|\delta U_1\|_3 + \|U\|_{k+1}\|\delta U_1\|_3\|\delta U_2\|_3\}.$$

2.2. Linearized equations at the origin and dispersion relation

The linearization at the origin of system (1.6), (1.7) leads to

(2.3) $\quad \mathcal{G}^{(0)}(\psi) - \mathbf{u} \cdot \nabla_X \eta = 0,$

(2.4) $\quad \mathbf{u} \cdot \nabla_X \psi + \mu \eta = 0,$

where the following operator
$$\mathcal{G}^{(0)} = (-\Delta)^{1/2}$$
is defined more precisely in Appendix A.2. Now expanding in Fourier series, we have
$$\psi(X) = \sum_{K \in \Gamma'} \psi_K e^{iK \cdot X}, \quad \eta(X) = \sum_{K \in \Gamma'} \eta_K e^{iK \cdot X},$$
and (2.3), (2.4) give for any $K \in \Gamma'$
$$|K|\psi_K - i(K \cdot \mathbf{u})\eta_K = 0,$$
$$i(K \cdot \mathbf{u})\psi_K + \mu\eta_K = 0.$$
Hence, the *dispersion relation* reads

(2.5) $$\Delta(K, \mu, \mathbf{u}) \stackrel{def}{=} \mu|K| - (K \cdot \mathbf{u})^2 = 0.$$

The point now to discuss is the number of solutions $K \in \Gamma'$ of (2.5), for a fixed vector $\mathbf{u} \in \mathbb{S}_1$, and a fixed parameter μ. We restrict our analysis to a *lattice Γ' generated by two vectors K_1 and K_2 symmetric with respect to the x_1- axis, taken in the direction of* $\mathbf{u}$, which is the situation if one is looking for *short crested waves*:
$$K_1 = (1, \tau), \quad K_2 = (1, -\tau)$$
where τ is positive. When τ is small, the lattice Γ of periods is formed with diamonds elongated in the x_2 direction (see Figure 1). Taking 1 for the first component of K_1 implies that we choose the length scale L as the wave length in the x_1- direction divided by 2π.

We consider in what follows, the cases when the direction $\mathbf{u}_0$ of the travelling waves at criticality is the x_1- axis, and the critical parameter $\mu_c = (1 + \tau^2)^{-1/2}$ is such that the equation for $(m_1, m_2) \in \mathbb{N}^2$

(2.6) $$\mu_c\sqrt{m_1^2 + \tau^2 m_2^2} - m_1^2 = 0$$

has only the solution
$$(m_1, m_2) = (1, 1).$$
In case we have a solution $(m_1, m_2) \neq (1, 1)$, one can make the change $(\mu_c, \tau) \mapsto (\frac{\mu_0}{m_1}, \tau\frac{m_2}{m_1})$ to recover the case we study here. Moreover, changing μ into μ/m_1 corresponds to changing the length scale L into L/m_1 which indeed corresponds to the new wave length in the x_1 direction. So, it is clear that *we do not restrict the generality* in choosing the case of a solution $(m_1, m_2) = (1, 1)$.

Notice that for any integer l, when $\tau = l$ or $1/l$, there is an infinite number of solutions (m_1, m_2) of (2.6), hence we need to avoid such choices for τ.

Remark. We notice here the fundamental difference between the present type of study and the works using spatial dynamics for finding travelling waves, as for instance Groves and Haragus in [**GH**]. Their study only consider cases with surface tension, and cannot work without surface tension, since this would lead to an infinite set of imaginary eigenvalues $\pm im_1$ (hence preventing the use of center manifold reduction), with no restriction for m_1 to be an integer, while $m_2 \in \mathbb{N}$ (this corresponds to fixing the length scale with the period in x_2, transverse to the direction of the travelling waves.

2.3. Formal computation of 3-dimensional waves

In this section we make a formal bifurcation analysis for the simple case. We denote by μ_c the critical value of μ, and we denote by $\mathbf{u}_0 = (1,0)$ the critical direction for the waves (this direction of propagation may be changed for bifurcating travelling waves). The lattice Γ' is generated by the two symmetric wave vectors $K_1 = (1, \tau), K_2 = (1, -\tau)$, where

$$\mu_c^{-2} = 1 + \tau^2.$$

We notice that we have the following Fourier series for $U = (\psi, \eta)$:

$$U = \sum_{n=(n_1,n_2)\in\mathbb{Z}^2} U_n e^{i(n_1 K_1 \cdot X + n_2 K_2 \cdot X)}, \quad U_n = (\psi_n, \eta_n), \quad \psi_0 = 0,$$

and we notice that

$$n_1 K_1 \cdot X + n_2 K_2 \cdot X = m_1 x_1 + \tau m_2 x_2,$$
$$m_1 = n_1 + n_2, \quad m_2 = n_1 - n_2,$$

which gives functions which are $2\pi-$ periodic in x_1 and $2\pi/\tau-$ periodic in x_2.

We already noticed the equivariance of system (1.6), (1.7) with respect to translations of the plane, represented by the linear operator $\mathcal{T}_\mathbf{v}$, $\mathbf{v}$ being any vector of the plane. Let us complete the symmetry properties of our system by the symmetries $\mathcal{S}_0$ and $\mathcal{S}_1$ defined by the representations of respectively the symmetry with respect to 0, and the symmetry with respect to x_1 axis

$$(2.7)\ \mathcal{S}_0 U = \sum_{n=(n_1,n_2)\in\mathbb{Z}^2} (SU_n) e^{-i(n_1 K_1 \cdot X + n_2 K_2 \cdot X)}, SU_n = (-\psi_n, \eta_n),$$

$$(2.8)\ \mathcal{S}_1 U = \sum_{n=(n_1,n_2)\in\mathbb{Z}^2} U_n e^{i(n_1 K_2 \cdot X + n_2 K_1 \cdot X)}.$$

The system (1.6), (1.7) is equivariant, under the symmetry $\mathcal{S}_0$ in all cases, while it is equivariant under $\mathcal{S}_1$ only if

$$\mathbf{u} \cdot K_1 = \mathbf{u} \cdot K_2.$$

In particular, in such a case we have

$$\mathcal{L}_0 \mathcal{S}_1 = \mathcal{S}_1 \mathcal{L}_0, \quad \mathcal{L}_0 \mathcal{S}_0 = \mathcal{S}_0 \mathcal{L}_0,$$

where we denote by $\mathcal{L}_0$ the symmetric linearized operator for $\mu = \mu_c$ and $\mathbf{u} = \mathbf{u}_0$

$$(2.9) \qquad \mathcal{L}_0 = \begin{pmatrix} \mathcal{G}^{(0)} & -\mathbf{u}_0 \cdot \nabla \\ \mathbf{u}_0 \cdot \nabla & \mu_c \end{pmatrix}.$$

Notice that the commutation property with the linear operator $\mathcal{S}_0$ is not trivial since the choice of writing our system in the moving frame selects the direction **u** which breaks a reflection symmetry. Indeed the symmetry property results from the galilean invariance of the Euler equations.

In the following Theorem, we use the two parameters $\tilde{\mu} = \mu - \mu_c$ and $\omega = \mathbf{u} - \mathbf{u}_0$ and we notice that, since **u** is unitary, we have

$$\omega = (\omega_1, \omega_2),$$
$$\omega_2 = \frac{1}{2\tau}\omega \cdot (K_1 - K_2),$$
$$\omega_1 = -\frac{\omega_2^2}{2} + O(\omega_2^4).$$

In the following sections we prove the following

THEOREM 2.3. *Assume we are in the simple case, i.e. for τ and the critical value of the parameter $\mu_c(\tau) = (1+\tau^2)^{-1/2} < 1$ such that the equation $n^2 + \tau^2 m^2 = \mu_c^{-2} n^4$ has only the solution $(n,m) = (1,1)$ in $\mathbb{N}^2$. Then, for any $N \geq 1$, and any $\mathbf{v} \in \mathbb{R}^2$, approximate 3-dimensional waves are given by $\mathcal{T}_\mathbf{v} U_\varepsilon^{(N)}$ (T^2− torus family of solutions) where $K_1 = (1, \tau)$, $K_2 = (1, -\tau)$ are the wave vectors, and*

$$U_\varepsilon^{(N)} = (\psi, \eta)_\varepsilon^{(N)} = \sum_{(p_1,p_2)\in\mathbb{N}^2,\, p_1+p_2\leq N} \varepsilon_1^{p_1}\varepsilon_2^{p_2} U^{(p_1,p_2)} \in \mathbb{H}^k, \text{ for any } k,$$

$$U^{(1,0)} = \xi_1 = (\sin(K_1 \cdot X), \frac{-1}{\mu_c}\cos(K_1 \cdot X)),$$

$$U^{(0,1)} = \xi_2 = (\sin(K_2 \cdot X), \frac{-1}{\mu_c}\cos(K_2 \cdot X)),$$

$$U^{(2,0)} = \left(\frac{-1}{2\mu_c^2}\sin(2K_1 \cdot X), \frac{1}{2\mu_c^3}\cos(2K_1 \cdot X)\right),$$

$$U^{(0,2)} = \left(\frac{-1}{2\mu_c^2}\sin(2K_2 \cdot X), \frac{1}{2\mu_c^3}\cos(2K_2 \cdot X)\right),$$

$$U^{(1,1)} = \left(\frac{1-2\mu_c}{\mu_c(2-\mu_c)}\sin((K_1+K_2)\cdot X), \frac{\mu_c^2+2\mu_c-2}{\mu_c^2(2-\mu_c)}\cos((K_1+K_2)\cdot X)\right) +$$
$$+ \left(0, \frac{\tau^2}{\mu_c}\cos((K_1-K_2)\cdot X)\right),$$

$$\tilde{\mu} = -\frac{\mu_c^2}{8}(\alpha_0 + \beta_0)(\varepsilon_1^2 + \varepsilon_2^2) + O\{(\varepsilon_1^2 + \varepsilon_2^2)^2\},$$
$$\omega \cdot (K_1 - K_2) = (\varepsilon_1^2 - \varepsilon_2^2)\left(\frac{\mu_c}{8}(\alpha_0 - \beta_0) + O\{(\varepsilon_1^2 + \varepsilon_2^2)\}\right),$$

with

$$\alpha_0 + \beta_0 = \frac{4}{\mu_c^2}\left(-\frac{1}{\mu_c^3} + \frac{2}{\mu_c^2} + \frac{3}{\mu_c} - 8 - 2\mu_c + \frac{9}{2-\mu_c}\right)$$

$$\beta_0 - \alpha_0 = \frac{4}{\mu_c^2}\left(-\frac{3}{\mu_c^3} + \frac{2}{\mu_c^2} + \frac{3}{\mu_c} - 8 - 2\mu_c + \frac{9}{2-\mu_c}\right),$$

and where for any N and k

$$\mathcal{F}(U_\varepsilon^{(N)}, \mu_c + \tilde{\mu}, \mathbf{u}_0 + \omega) = (\varepsilon_1^2 + \varepsilon_2^2)^{\frac{N+1}{2}} Q_\varepsilon,$$

Q_ε uniformly bounded in $\mathbb{H}^k$, with respect to $\varepsilon_1, \varepsilon_2$. There are critical values τ_c and τ_c' of τ such that $(\alpha_0 + \beta_0)(\tau_c) = 0$, and $(\alpha_0 - \beta_0)(\tau_c') = 0$, and $\alpha_0 + \beta_0$ is positive for $\tau \in (0, \tau_c)$, negative for $\tau > \tau_c$, $\tau_c \approx 2.48$, while $\alpha_0 - \beta_0$ is negative for $\tau \in (0, \tau_c')$, positive for $\tau > \tau_c'$, $\tau_c' \approx 0.504$. Moreover, for $\varepsilon_1 = \varepsilon_2$ we have "diamond waves" where the direction of propagation $\mathbf{u}$ is along the x_1– axis; and for $\varepsilon_2 = 0$ (resp. $\varepsilon_1 = 0$) we obtain 2-dimensional travelling waves of wave vector K_1 (resp. K_2). All solutions are invariant under the shift $\mathcal{T}_{\mathbf{v}_0}: X \mapsto X + (\pi, \pi/\tau)$ and $\psi_\varepsilon^{(N)}$ is odd in X, while $\eta_\varepsilon^{(N)}$ is even in X.

Remark 1: In this Theorem we assume that equation (2.6) for $(m_1, m_2) \in \mathbb{Z}^2$ has only the four solutions $(m_1, m_2) = (\pm 1, \pm 1)$, corresponding to the four wave vectors $K = \pm K_1$ and $\pm K_2$. The corresponding pattern of the waves for $\varepsilon_1 = \varepsilon_2$ is in diamond form, and for τ close to 0, the diamonds are flattened in the x_1 direction, and elongated in x_2 direction, looking like flattened hexagons or flattened rectangles, because of the elongated shape of crests and troughs. This last case is indeed observed experimentally for deep fluid layers. Nearly all (in the measure sense) values of τ are indeed such that we are in the simple case.

Remark 2: The above Theorem is stated differently in Theorem 4.1 of [**CrN1**] in the case with surface tension ; indeed we prove here that the manifold of solutions has a simple formulation in terms of the two parameters, and this result trivially extends in the case with surface tension.

Remark 3: Since $\mu = gL/c^2$, the result of Theorem 2.3 about the sign of $\alpha_0 + \beta_0$ shows that for $\tau < \tau_c$ the bifurcation of "diamond waves" (i.e. $\omega = 0$) occurs for a velocity c of the waves larger than the critical velocity c_0 corresponding to μ_c, while for $\tau > \tau_c$ the bifurcation occurs for $c < c_0$. This is in accordance with the numerical results of Bridges et al [**BDM**] (see p. 166-167 with $A_1 = A_2$ real, $T_{10} = 1, \tau = 0$ (no surface tension)). Notice that $\tau_c^2 \approx 6.15$, i.e. this corresponds to a critical angle θ between the wave number K_1 and the direction of the travelling wave, such that $\theta \approx 68^0$, which is very large, and not easy to reach experimentally.

Remark 4: Notice that for τ near τ_c' we still have "diamond waves' (even for $\tau = \tau_c'$) and it may exist other bifurcating 3-dimensional waves, as noticed in [**BDM**]. However to confirm this, we need to compute at least coefficients of order 4 in (B.5).

Remark 5: Experiments made by Hammack et al [**HHS**] correspond to diamond waves ($\varepsilon_1 = \varepsilon_2 = \varepsilon$). Looking at the Fourier spectrum (see [**HHS**] figures 6, 7, 8) of the elevation η of the waves, we see clearly the modes K_1, K_2, $K_1 + K_2$, $2K_1$, $2K_2$. These experiments are made with an amplitude ε small enough so that it implies that we only see Fourier modes correspondind to orders ε and ε^2. An a priori strange fact is that the Fourier modes $K_1 - K_2$ and $K_2 - K_1$ are not present in the experimental spectra, even for larger amplitudes as in figure 9. The explanation is clear in looking at the expression of $U^{(1,1)}$ in Theorem 2.3 where we observe that the amplitude of the modes $K_1 - K_2$ and $K_2 - K_1$ are of order τ^2, contrary to other components, and we observe that in the above experiments $\tau = \tan\theta$ is very small since the angle θ is less than 10 degrees.

Proof: the proof is made in Appendix B.

2.4. Geometric pattern of diamond waves

Diamond waves are obtained for $\varepsilon_1 = \varepsilon_2$, they propagate along the x_1-axis, and possess the symmetry $\mathcal{S}_1$. In making $\varepsilon_1 = \varepsilon_2 = \varepsilon/2$ in Theorem 2.3 we obtain

$$(2.10) \quad U_\varepsilon^{(N)} = (\psi, \eta)_\varepsilon^{(N)} = \sum_{n \in \mathbb{N}, n \leq N} \varepsilon^n U^{(n)} \in \mathbb{H}^k, \text{ for any } k,$$

$$U^{(1)} = \xi_0 = (\sin x_1 \cos \tau x_2, -\frac{1}{\mu_c} \cos x_1 \cos \tau x_2),$$

$$U^{(2)} = \frac{1}{4}(U^{(2,0)} + U^{(0,2)} + U^{(1,1)}),$$

where $\psi_\varepsilon^{(N)} \in H_{o,e}^k$, $\eta_\varepsilon^{(N)} \in H_{e,e}^k$ meaning that $\psi_\varepsilon^{(N)}$ is odd in x_1, even in x_2, and $\eta_\varepsilon^{(N)}$ is even in both coordinates. Moreover, we have

$$\mu - \mu_c = \tilde{\mu} = \varepsilon^2 \mu_1 + O(\varepsilon^4)$$

with

$$(2.11) \quad \begin{aligned} \mu_c &= (1+\tau^2)^{-1/2}, \\ \mu_1 &= -\frac{\mu_c^2}{16}(\alpha_0 + \beta_0) \\ &= \frac{1}{4\mu_c^3} - \frac{1}{2\mu_c^2} - \frac{3}{4\mu_c} + 2 + \frac{\mu_c}{2} - \frac{9}{4(2-\mu_c)}. \end{aligned}$$

For τ close to 0 (corresponds to some of the experiments shown in [**HHS**]), one has

$$\eta = \tilde{\varepsilon} \cos x_1 \cos \tau x_2 + \frac{\tilde{\varepsilon}^2}{4} \cos 2x_1 (1 + \cos 2\tau x_2) + O(\tilde{\varepsilon}^2 \tau^2 + |\tilde{\varepsilon}|^3)$$

with

$$\tilde{\varepsilon} = -\varepsilon/\mu_c \sim -\varepsilon.$$

One can assume that $\tilde{\varepsilon} > 0$ since $\tilde{\varepsilon} < 0$ would correspond to a shift by π of x_1. Then the above formula shows that crests (maxima) and troughs (minima) are elongated in the x_2 direction, with crests sharper than the troughs, and there are "nodal" lines ($\eta \approx 0$) (as noticed in experiments [**HHS**]) at

$$x_2 = \pi/2\tau + n\pi/\tau, \quad n \in \mathbb{Z},$$

where η is of order $O(\tilde{\varepsilon}^2\tau^2 + |\tilde{\varepsilon}|^3)$. So, the pattern roughly looks asymptotically like *rectangles elongated in x_2 direction, narrow around the crests, wide around the troughs, organized in staggered rows*. Notice that when $\tau \to 0$, we have $\mu_1 \sim -3/4$, hence $\varepsilon^2 \sim (4/3)(\mu_c - \mu)$, i.e.

$$\frac{c-c_0}{c_0} \sim \frac{3}{8}F_0^2\tilde{\varepsilon}^2, \quad F_0 = \frac{c_0}{\sqrt{gL}},$$

where F_0 is the Froude number built with the short wave length L.

For larger, still small, values of τ, the nodal lines disappear and *the pattern looks like hexagons* where two sides of crests, parallel to the x_2 axis, are connected to the nearest tip of two analogue crests, shifted by half of the wave length in x_1 and x_2 directions.

For values of τ near 1, the pattern of the surface looks like juxtaposition of squares.

For large τ, i.e. in particular $\tau > \tau_c$, we have for the free surface truncated at order ε^2

$$\eta = \tilde{\varepsilon}\cos x_1 \cos\tau x_2 + \frac{\tau\tilde{\varepsilon}^2}{4}\cos 2\tau x_2(1 + \cos 2x_1) + O(\tilde{\varepsilon}^2)$$

where

$$\tilde{\varepsilon} = -\varepsilon/\mu_c \sim -\varepsilon\tau.$$

The above formula shows that crests (maxima) and troughs (minima) are elongated in the x_1 direction, with crests sharper than the troughs, and there are "nodal" lines ($\eta \approx 0$) at

$$x_1 = \pi/2 + n\pi, \quad n \in \mathbb{Z}.$$

Moreover, in the above formula, we see that τ is in factor of $\tilde{\varepsilon}^2$, which means that the second order term in the expansion influences much sooner the shape of the surface as $\tilde{\varepsilon}$ increases. In particular for small values of $\tilde{\varepsilon}$ there are local maxima between two minima in the troughs. This phenomenon is seen in experiments (see [**HHS**]) in the case when τ is small, however the values of $\tilde{\varepsilon}$ allowing such a phenomenon for τ small are $O(1)$ and cannot be justified mathematically. So, when τ is large, the pattern roughly looks asymptotically like *rectangles elongated in x_1 direction, narrow around the crests, wide around the troughs, organized in staggered rows, and where local maxima in the middle of the troughs may occur for a large enough amplitude*. Notice in addition that when $\tau \to \infty$, then $\mu_1 \sim \frac{1}{4}\tau^{7/2}$, hence $\varepsilon^2\tau^2 \sim \frac{4}{\tau^{3/2}}(\mu - \mu_c)$, i.e.

$$\frac{c_0 - c}{c_0} \sim \frac{1}{8}\tau^{3/2}F_0^2\tilde{\varepsilon}^2, \quad F_0 = \frac{c_0}{\sqrt{gL}},$$

where we observe that in this last formula L is the physical wave length along x_1, which is in this case the long wave length ($= \tau L_1$, if L_1 denotes the short one). We plot at Figure 1 the elevation $\eta_\varepsilon^{(2)}$ a) for $\tau = 1/5$ ($\theta \sim 11.3^o$), $\varepsilon = 0.8\mu_c$, b) for $\tau = 1/2$ ($\theta \sim 26.5^o$), $\varepsilon = 0.6\mu_c$. These cases correspond to τ very small or moderately small, currently observed in experiments (see [**HHS**]). Observe however that in both cases we need to consider τ not exactly $1/5$ or $1/2$ since both cases are not "simple cases" as required at Theorem 2.3. Indeed, if we consider solutions $(n,m) \in \mathbb{N}^2$ different from $(1,1)$ for the critical dispersion relation

$$(2.12) \qquad n^2 + \tau^2 m^2 = (1+\tau^2)n^4,$$

then the smallest values are $(n,m) = (61, 18971)$ for $\tau = 1/5$, and $(13, 377)$ for $\tau = 1/2$. This means that the computation fails for coefficients of ε^{18971} in the first case, and ε^{377} in the second case. Notice that the formal computations made by Roberts and Schwartz [**RoSc**] (1983) correspond to diamond waves with $\tau = 1$ ($\theta = 45^o$) and $\tau = \sqrt{3}$ ($\theta = 60^o$). They observed numerically the flattening of troughs and sharpening of crests. However, both cases are not "simple cases" in the sense of Theorem 2.3, and these formal computation should break at order ε^{35} for $\tau = 1$, and at order ε^{195} for $\tau = \sqrt{3}$ (due to solutions of (2.12) $(n,m) = (5,35)$ for $\tau = 1$, and $(n,m) = (13,195)$ for $\tau = \sqrt{3}$).

CHAPTER 3

Linearized Operator

In this chapter we study the linearized problem at a non zero $U = (\psi, \eta)$ for the system (1.6), (1.7). We restrict our study to diamond waves, i.e. the direction of the waves we are looking for is $\mathbf{u} = \mathbf{u}_0 = (1, 0)$ and the system possesses the symmetries $\mathcal{S}_0$ and $\mathcal{S}_1$ (see (2.7), (2.8)). The purpose is to invert the linearized operator, for being able to use the Newton method, as it is required in the Nash-Moser theorem.

3.1. Linearized system in $(\psi, \eta) \neq 0$

Let us write the nonlinear system (1.6), (1.7) under the form (1.5)

$$\mathcal{F}(U, \mu) = 0,$$

where

$$U = (\psi, \eta),$$

and we omit the argument $\mathbf{u}_0$ since it is now fixed. Then, for any given (f, g) the linear system

$$\partial_U \mathcal{F}(U, \mu)[\delta U] = F, \quad F := (f, g)$$

can be written as follows

$$\partial_\eta \mathcal{G}_\eta[\delta \eta]\psi + \mathcal{G}_\eta(\delta \psi) - \mathbf{u}_0 \cdot \nabla(\delta \eta) = f,$$

$$V \cdot \nabla(\delta \psi) + \mu \delta \eta + \left(\mathfrak{b}^2 \nabla \eta - \mathfrak{b}(\nabla \psi + \mathbf{u}_0) \right) \cdot \nabla(\delta \eta) = g$$

where

(3.1) $\quad V = \nabla \psi + \mathbf{u}_0 - \mathfrak{b} \nabla \eta, \quad \mathfrak{b} = \dfrac{1}{1 + |\nabla \eta|^2} \{ \nabla \eta \cdot (\mathbf{u}_0 + \nabla \psi) \}.$

Now defining

$$\delta \phi = \delta \psi - \mathfrak{b} \delta \eta,$$

and after using (2.1,2.2), we obtain the new system

(3.2) $\quad \mathcal{L}(U, \mu)[\delta \phi, \delta \eta] = F + \mathcal{R}(\mathcal{F}, U)[\delta U],$

where the linear *symmetric* operator $\mathcal{L}(U, \mu)$ is defined by

(3.3) $\quad \mathcal{L}(U, \mu) = \begin{pmatrix} \mathcal{G}_\eta & \mathcal{J}^* \\ \mathcal{J} & \mathfrak{a} \end{pmatrix},$

(3.4) $\quad \mathcal{J} = V \cdot \nabla(\cdot), \quad \mathfrak{a} = V \cdot \nabla \mathfrak{b} + \mu.$

The rest $\mathcal{R}$ has the form

$$\mathcal{R}(\mathcal{F},U)[\delta U] = (R_1(\mathcal{F},U)[\delta U], 0),$$

$$R_1(\mathcal{F},U)[\delta U] = \mathcal{G}_\eta\left(\frac{\mathcal{F}_1 \delta \eta}{1+(\nabla\eta)^2}\right) - \nabla \cdot \left(\frac{\mathcal{F}_1 \delta \eta}{1+(\nabla\eta)^2}\nabla\eta\right),$$

and cancels when U is a solution of $\mathcal{F}(U,\mu) = 0$. We also notice that for $U = (\psi,\eta) \in \mathbb{H}_{(S)}^m$ where

$$\mathbb{H}_{(S)}^m = \{U \in \mathbb{H}^m(\mathbb{R}^2/\Gamma) : \psi \text{ odd in } x_1, \text{ even in } x_2, \eta \text{ even in } x_1 \text{ and in } x_2\},$$

then

$$V = (V_1, V_2) \in H_{e,e}^{m-1}(\mathbb{R}^2/\Gamma) \times H_{o,o}^{m-1}(\mathbb{R}^2/\Gamma),$$

$$\mathfrak{b} \in H_{o,e}^{m-1}(\mathbb{R}^2/\Gamma), \quad \mathfrak{a} \in H_{e,e}^{m-2}(\mathbb{R}^2/\Gamma),$$

all these functions being invariant under the shift

$$\mathcal{T}_{\mathbf{v}_0} : (x_1, x_2) \mapsto (x_1 + \pi, x_2 + \pi/\tau).$$

Moreover we have the following "tame" estimates

LEMMA 3.1. *Let $U \in \mathbb{H}_{(S)}^m$, $m \geq 3$. Then, there exists $M_3 > 0$ such that for $\|U\|_3 \leq M_3$, one has*

$$\|V - \mathbf{u}\|_{m-1} + \|\mathfrak{a} - \mu\|_{m-2} \leq c_m(M_3)\|U\|_m,$$

$$\|R_1(\mathcal{F},U)[\delta U]\|_{m-2} \leq c_s(M_3)\{\|\mathcal{F}_1\|_2(\|\eta\|_m\|\delta\eta\|_2 + \|\delta\eta\|_{m-1}) + \|\mathcal{F}_1\|_{m-1}\|\delta\eta\|_2\}.$$

PROOF. The tame estimates on $V - \mathbf{u}$ and $\mathfrak{a} - \mu$ result directly from their definitions, from the following inequality, valid for any $f,g \in H^m(\mathbb{R}^2/\Gamma)$, $m \geq 2$

(3.5) $$\|fg\|_m \leq c_m\{\|f\|_2\|g\|_m + \|f\|_m\|g\|_2\},$$

and from interpolation estimates like

(3.6) $$\|f\|_{\lambda\alpha + (1-\lambda)\beta} \leq c\|f\|_\alpha^\lambda \|f\|_\beta^{1-\lambda},$$

which leads to

$$\|f\|_{\alpha_2}\|f\|_{\beta_2} \leq c\|f\|_{\alpha_1}\|f\|_{\beta_1}$$

when

$$0 \leq \alpha_1 \leq \alpha_2 \leq \beta_2 \leq \beta_1, \quad \alpha_1 + \beta_1 = \alpha_2 + \beta_2.$$

The tame estimate on $R_1(\mathcal{F},U)[\delta U]$ follows from the tame estimates (3.5), (1.9) and from the following interpolation estimate deduced from (3.6):

$$\|f\|_3\|g\|_{k+1} \leq c_k\{\|f\|_2\|g\|_{k+2} + \|f\|_{k+1}\|g\|_3\}.$$

□

The main problem in using the Nash Moser theorem, is to invert the approximate linearized system, i.e. invert the linear system

$$\mathcal{L}[\delta\phi, \delta\eta] = (f, g),$$

which leads to the scalar equation

(3.7) $$\mathcal{G}_\eta(\delta\phi) - \mathcal{J}^*(\frac{1}{\mathfrak{a}}\mathcal{J}(\delta\phi)) = h$$

with

$$h = f - \mathcal{J}^*(\frac{1}{\mathfrak{a}}g) \in H^s_{o,e}(\mathbb{R}^2/\Gamma)$$

and where we look for $\delta\phi$ in some $H^{s-r}_{o,e}(\mathbb{R}^2/\Gamma)$.

3.2. Pseudodifferential operators and diffeomorphism of the torus

In this subsection we use a diffeomorphism: $\mathbb{R}^2 \to \mathbb{R}^2$, such that the principal part of the symbol of the linear operator occurring in (3.7) has a simplified structure. Its new structure will allow us to use further a suitable descent method for obtaining, at the end of the process, a pseudodifferential operator equation with constant coefficients, plus a perturbation operator of "small" order.

Let us denote the change of coordinates by $X = X(Y)$, where $X(\cdot)$ is a not yet determined diffeomorphism of $\mathbb{R}^2$ such that

(3.8) $$X(Y) = \mathbb{T}Y + \widetilde{\mathcal{V}}(Y), \quad \mathbb{T}Y = (y_1, y_2/\tau),$$

We assume that $\widetilde{\mathcal{V}}(Y)$ and $\tilde{\eta}(Y) = \eta(X(Y))$ satisfy the following

CONDITION 3.2. *Functions $\tilde{\eta}$ and $\widetilde{\mathcal{V}}$ are doubly 2π-periodic, $\tilde{\eta}$ is even in y_1 and y_2, $\widetilde{\mathcal{V}}_1$ is odd in y_1 and even in y_2, $\widetilde{\mathcal{V}}_2$ is odd in y_2 and even in y_1, and*

$$\tilde{\eta}(y_1 + \pi, y_2 + \pi) = \tilde{\eta}(y_1, y_2), \quad \widetilde{\mathcal{V}}(y_1 + \pi, y_2 + \pi) = \widetilde{\mathcal{V}}(y_1, y_2).$$

In particular, $X(Y)$ takes diffeomorphically $\mathbb{R}^2/\Gamma_1$ onto $\mathbb{R}^2/\Gamma_\tau$. The lattices of periods Γ_1 and Γ_τ are respectively the dual of lattices Γ'_1 and Γ'_τ generated by the wave vectors $(1, \pm 1)$ and $(1, \pm \tau)$. In new coordinates the free surface has the parametric representation

$$x = \mathbf{r}(Y) := (X_1(Y), X_2(Y), \widetilde{\eta}(Y))^t,$$

with 2x2 matrix $\mathbb{G}(Y)$ of the *first fundamental form of the free surface* defined by $g_{ij} = \partial_{y_i}\mathbf{r} \cdot \partial_{y_j}\mathbf{r}$. We denote by J the determinant of the Jacobian matrix $\mathbb{B}(Y) = \nabla_Y X(Y)$.

Our aim is to simplify the structure of the operators involved in the basic equation (3.7) by choosing an appropriate change of coordinates. The most suitable tool for organizing such a process is the theory of pseudodifferential operators, and we begin with recalling the definition of a pseudodifferential operator. We consider the class of integro-differential operators on a two-dimensional torus having the representation

$$\mathfrak{A}u(Y) = \frac{1}{2\pi} \sum_{k \in \mathbb{Z}^2} e^{ikY} A(Y,k)\widehat{u}(k), \quad \widehat{u}(k) = \frac{1}{2\pi} \int_{T^2} e^{-ikY} u(Y) dY,$$

which properties are completely characterised by the function $A : \mathbf{T}^2 \times \mathbb{R}^2 \mapsto \mathbb{C}$ named the symbol of $\mathfrak{A}$. We say that $\mathfrak{A}$ is a pseudodifferential operator, if its symbol satisfy the condition

CONDITION 3.3. *There are integers $l > 0$, $m \geq 0$ and a real r named the order of the operator $\mathfrak{A}$ so that*

$$|\mathfrak{A}|^r_{m,l} = \|A(\cdot,0)\|_{C^l} + \sup_{k \in \mathbb{Z}^2 \setminus \{0\}} \sup_{|\alpha| \leq m} |k|^{|\alpha|-r} \|\partial^\alpha_k A(\cdot,k)\|_{C^l} < \infty.$$

Pseudodifferential operators enjoy many remarkable properties including explicit formulae for compositions and commutators (see Appendix F for references and more details). Important examples of such operators are the first-order pseudodifferential operator $\mathcal{G}^{(0)} = (-\Delta)^{1/2}$ with the symbol $|\mathbb{T}^{-1}k|$ and the second-order pseudodifferential operator

$$(3.9) \qquad \mathfrak{L} = \nu \mathfrak{D}_1^2 + (-\Delta)^{1/2}, \text{ where } \mathfrak{D}_1 = \partial_{y_1},$$

with the symbol

$$(3.10) \qquad L(k) = -\nu k_1^2 + |\mathbb{T}^{-1}k|.$$

On the other hand, integro-differential operators $\mathfrak{D}_1^j$, defined by

$$(3.11) \qquad \mathfrak{D}_1^j u(Y) = \frac{1}{2\pi} \sum_{k_1 \neq 0} e^{ikY} \left(ik_1\right)^j \widehat{u}_k, \quad j \in \mathbb{Z}.$$

are not pseudodifferential for $j \leq 0$, which easy follows from the formulae

$$\mathfrak{D}_1^j u = \partial^j_{y_1} u \text{ for } j > 0, \text{ and } \mathfrak{D}_1^0 u = \Pi_1 u = u - \frac{1}{2\pi} \int_{-\pi}^{\pi} u(s, y_2) ds.$$

Further we will consider also the special class of zero-order pseudodifferential operators $\mathfrak{A}$ with symbols having the form of composition $A(Y, \xi(k))$, where the vector field $\xi(k) = (\xi_1(k), \xi_2(k))$ is defined by

$$\xi(k) = \mathbb{T}^{-1}k/|\mathbb{T}^{-1}k| \text{ for } k \neq 0, \quad \xi(0) = 0.$$

The metric properties of such operators are characterized by the norm

$$(3.12) \qquad |\mathfrak{A}|_{m,l} = \sup_{|\alpha| \leq m} \sup_{|\xi| \leq 1} \|\partial^\alpha_\xi A(\cdot,\xi)\|_{C^l} < \infty,$$

which is equivalent to the norm $|\mathfrak{A}|^0_{m,l}$. By abuse of notation, further we will write simply ξ instead of $\xi(k)$ and use both the notations $A(Y, \xi_1, \xi_2)$ and $A(Y, \xi)$ for $A(Y, \xi)$.

We are now in a position to formulate the main result of this section.

THEOREM 3.4. *For any integers $\rho, m \geq 14$, real $\tau \in (\delta, \delta^{-1})$, and $U = (\psi, \eta) \in \mathbb{H}^m_{(S)}(\mathbb{R}^2/\Gamma_\tau)$, there exists $\varepsilon_0 > 0$ so that for*

$$\|U\|_\rho \leq \varepsilon \text{ with } \varepsilon \in [0, \varepsilon_0]$$

there are a diffeomorphism of the torus of the form (3.8), satisfying Condition 3.2, zero-order pseudodifferential operators $\mathfrak{A}$, $\mathfrak{B}$ and an integro-differential operator $\mathfrak{L}_{-1}$ of order -1 such that:

(i) *The identity*

$$(3.13) \quad \left[\mathcal{G}_\eta \check{u} - \mathcal{J}^*\left(\frac{1}{\mathfrak{a}}\mathcal{J}\check{u}\right)\right] \circ (\mathbb{T} + \tilde{\mathcal{V}}) = \kappa\left[\mathfrak{L}u + \mathfrak{A}\mathfrak{D}_1 u + \mathfrak{B}u + \mathfrak{L}_{-1}u\right]$$

holds true for any $u \in H^2_{o,e}(\mathbb{R}^2/\Gamma_1)$ and $\check{u}(X) = u(Y(X))$.

(ii) *The operators $\mathfrak{A}$, $\mathfrak{B}$ and $\mathfrak{L}_{-1}$ have the bounds*

$$|\mathfrak{A}|_{4,m-6} + |\mathfrak{B}|_{4,m-6} \leq c_m(\|U\|_6)\|U\|_m,$$
$$\|\mathfrak{L}_{-1}u\|_r \leq c\varepsilon\|u\|_{r-1}, \text{ for } 1 \leq r \leq \rho - 13,$$
$$\|\mathfrak{L}_{-1}u\|_s \leq c(\varepsilon\|u\|_{s-1} + \|U\|_{s+13}\|u\|_0).$$

(iii) *Operators $\mathfrak{A}$, $\mathfrak{B}$ and $\mathfrak{L}_{-1}$ are invariant with respect to the symmetries $Y \to \pm Y^*$, $Y^* = (-y_1, y_2)$ which is equivalent to the equivariant property*

$$(3.14) \quad \begin{array}{ll} \mathfrak{A}\mathfrak{D}_1 u(\pm Y^*) = \mathfrak{A}\mathfrak{D}_1 u^*(\pm Y), & \mathfrak{B}u(\pm Y^*) = \mathfrak{B}u^*(\pm Y), \\ \mathfrak{L}_{-1}u(\pm Y^*) = \mathfrak{L}_{-1}u^*(\pm Y), & u^*(Y) = u(Y^*); \end{array}$$

they are also invariant with respect to transform $Y \to Y + (\pi, \pi)$.

(iv) *Diffeomorphism (3.8) of the torus can be inverted as $Y = \mathbb{T}^{-1}(X - \mathcal{V}(X))$,*

$$y_1 = x_1 + d(x_1, x_2), \quad y_2 = \tau x_2 + \tau e(x_1, x_2).$$

Functions $d \in C^{m-4}_{o,e}(\mathbb{R}^2/\Gamma_\tau)$, $e \in C^{m-4}_{e,o}(\mathbb{R}^2/\Gamma_\tau)$, $\kappa, J \in C^{m-4}_{e,e}(\mathbb{R}^2/\Gamma_1)$ and parameter ν satisfy the inequalities

$$(3.15) \quad \|d\|_{C^{m-4}} + \|e\|_{C^{m-4}} \leq c_m(\|U\|_4)\|U\|_m, \quad |\nu - 1/\mu| \leq c(\|U\|_4)\|U\|_4,$$
$$(3.16) \quad \|\kappa - 1\|_{C^{m-5}} + \|J - 1/\tau\|_{C^{m-5}} \leq c_m(\|U\|_5)\|U\|_m.$$

The proof is based on two propositions, the first of which gives the representation of the Dirichlet-Neumann operator in the invariant parametric form, and the second shows that trajectories of liquid particles on the free surface forms a foliation of the two-dimensional torus.

In order to formulate them it is convenient to introduce the notations

$$\mathbf{G}_1(Y, k) = \sqrt{\mathbb{G}^{-1}k \cdot k}, \quad \mathbf{div}\,\mathbf{q}(Y) = \frac{1}{\sqrt{\det \mathbb{G}}}\,\mathrm{div}_Y\left(\sqrt{\det \mathbb{G}}\mathbf{q}(Y)\right).$$

Recall that $\mathbb{G}k \cdot k$ is the first fundamental form of the surface Σ.

THEOREM 3.5. *Suppose that functions $\tilde{\eta}$ and $\widetilde{\mathcal{V}}$ satisfy Condition 3.2 and there are integers ρ, l such that*

$$||\tilde{\eta}||_{C^\rho} + ||\widetilde{\mathcal{V}}||_{C^\rho} \leq \varepsilon, \quad 10 \leq \rho \leq l,$$
$$||\tilde{\eta}||_{C^s} + ||\widetilde{\mathcal{V}}||_{C^s} \leq E_l, \quad s \leq l.$$

Then there exists $\varepsilon_0 > 0$ depending on ρ and l only such that for $0 \leq \varepsilon \leq \varepsilon_0$ and 2π-periodic sufficiently smooth function u, the operator $\mathcal{G}_\eta$ has the representation

$$(3.17) \qquad \mathcal{G}_\eta \check{u} \circ (\mathbb{T} + \widetilde{\mathcal{V}}) = \mathcal{G}_1 u + \mathcal{G}_0 u + \mathcal{G}_{-1} u, \quad \check{u}(X) = u(Y(X)).$$

Here $\mathcal{G}_1$ is a first order pseudodifferential operator with symbol

$$(3.18) \qquad G_1(Y, k) = \frac{\sqrt{\det \mathbb{G}}}{J} \mathbf{G}_1(Y, k), \quad Y \in \mathbb{R}^2, \quad k \in \mathbb{Z}^2,$$

$\mathcal{G}_0$ *is a zero order pseudodifferential operator with symbol*

$$G_0 = \operatorname{Re} G_0 + i \operatorname{Im} G_0,$$

$$(3.19) \qquad \operatorname{Re} G_0(Y, k) = \frac{\det \mathbb{G}}{2J^2} \Big[\frac{1}{\mathbf{G}_1^2} Q(Y, k) + \mathbf{div}(\mathbb{G}^{-1} \nabla_Y \tilde{\eta}) \Big],$$

$$(3.20) \qquad \operatorname{Im} G_0(Y, k) = -\frac{\sqrt{\det \mathbb{G}}}{2J} \mathbf{div}(\nabla_k \mathbf{G}_1).$$

Here the quadratic form $Q(Y, .)$ is given by

$$(3.21) \quad Q(Y, k) = \frac{1}{2} \nabla_Y(\mathbb{G}^{-1} k \cdot k) \cdot (\mathbb{G}^{-1} \nabla_Y \tilde{\eta}) - \mathbb{G}^{-1} k \cdot \nabla_Y(\mathbb{G}^{-1} \nabla_Y \tilde{\eta} \cdot k),$$

and the operator $\mathcal{G}_0$ satisfies the estimates

$$(3.22) \qquad |\mathcal{G}_0 u|^0_{4, \rho-2} \leq c\varepsilon, \quad |\mathcal{G}_0 u|^0_{4, l-2} \leq c E_l,$$

while the linear operator $\mathcal{G}_{-1}$ satisfies

$$(3.23) \quad \begin{aligned} ||\mathcal{G}_{-1} u||_r &\leq c\varepsilon ||u||_{r-1}, \quad \text{for } 1 \leq r \leq \rho - 9, \\ ||\mathcal{G}_{-1} u||_s &\leq c(\varepsilon ||u||_{s-1} + E_l ||u||_0), \quad \text{for } s \leq l - 9. \end{aligned}$$

Moreover, operators $\mathcal{G}_1$, $\mathcal{G}_0$ and $\mathcal{G}_{-1}$ satisfy the symmetry properties

$$\mathcal{G}_j u(\pm Y^*) = \mathcal{G}_j u^*(\pm Y), \quad j = 1, 0, -1, \quad u^*(Y) = u(Y^*).$$

PROOF. The proof is given in Appendix G. □

The formula for the principal term G_1 is a classic result of the theory of pseudodifferential operators [**Ho**]. The expression for the second term in local Riemann coordinates was given in [**ABB**]. It seems that the general formulae (3.19), (3.20) are new. Note that the ratio $\det \mathbb{G}/J^2 = 1 + |\nabla_X \eta(X)|^2$ is a scalar invariant and the real part of G_0 can be rewritten in the invariant form

$$(3.24) \qquad \frac{J}{\sqrt{\det \mathbb{G}}} \operatorname{Re} G_0 = \frac{1}{2} \frac{LG - 2MF + NF}{EG - F^2} - \frac{1}{2} \frac{L\xi_1^2 + 2M\xi_1\xi_2 + N\xi_2^2}{E\xi_1^2 + 2F\xi_1\xi_2 + G\xi_2^2}.$$

Here we use the standard notations for the second fundamental form $L\xi_1^2 + 2M\xi_1\xi_2 + N\xi_2^2$ and the first fundamental form $E\xi_1^2 + 2F\xi_1\xi_2 + G\xi_2^2$ of the surface $x_3 = \eta(X)$, the vector ξ is connected with the covector k by the relation $k = \mathbb{G}\xi$. The right hand side of (3.24) is the difference between the mean curvature of Σ and half of the normal curvature of Σ in the direction ξ. Note also that the conclusion of Theorem 3.5 holds true without assumption on the smallness of ε, but the proof becomes more complicated and goes far beyond the scope of the paper.

LEMMA 3.6. *For $m \geq 4$, and $U \in \mathbb{H}^m_{(S)}(\mathbb{R}^2/\Gamma_\tau)$ with $\|U\|_4$ small enough, there exists a unique function $\mathcal{Z} \in \mathcal{C}^{m-3}(\mathbb{R}^2)$ such that*

$$(3.25) \qquad \frac{\partial \mathcal{Z}}{\partial z_1} = \frac{V_2}{V_1}(z_1, \mathcal{Z}), \quad \Pi_1 \mathcal{Z} = z_2,$$

where Π_1 denotes the average over a period in z_1, and V_i are the components of the vector field V defined by (3.1). Moreover, $\mathcal{Z}$ is even in z_1, odd in z_2,

$$\mathcal{Z}(Z) = \mathcal{Z}(Z + (2\pi, 0)) = \mathcal{Z}(Z + (0, 2\pi/\tau)) - 2\pi/\tau = \mathcal{Z}(Z + (\pi, \pi/\tau)) - \pi/\tau,$$

and the shifted function $T_\delta \mathcal{Z} = \mathcal{Z}(\cdot + \delta, \cdot)$ is solution of the system (3.25) where $V = V(\cdot + \delta, \cdot)$. Moreover, the mapping

$$(3.26) \qquad x_1 = z_1, \quad x_2 = z_2 - \widetilde{d}_1(z_1, z_2) = \mathcal{Z}(z_1, z_2),$$

with its inverse

$$z_1 = x_1, \quad z_2 = x_2 + d_1(x_1, x_2)$$

define automorphisms of the torus $\mathbb{R}/\Gamma_\tau$: $X \mapsto Z = \mathcal{U}_1(X)$, $X = \mathcal{U}_1^{-1}(Z)$. The functions d_1 and $\widetilde{d}_1$ have symmetry (e, o), as above and we have the following tame estimates

$$\|d_1\|_{C^{m-3}} + \|\widetilde{d}_1\|_{C^{m-3}} \leq c_m(\|U\|_4)\|U\|_m.$$

The automorphism $\mathcal{U}_1$ takes integral curves of the vector field V, which coincide with the bicharacteristics of the operator $\mathcal{G}_\eta - \mathcal{J}^*(\mathfrak{a}^{-1}\mathcal{J}\cdot)$, onto straight lines $\{z_2 = \text{const.}\}$. In other words, bicharacteristics form a foliation of the torus with a rotation number equal to 0.

PROOF. The proof is done in Appendix C. □

Let us turn to the proof of Theorem 3.4. We look for the desired diffeomorphism $Y \to X$ in the form of the composition $(\mathcal{U}_2 \circ \mathcal{U}_1)^{-1}$,

$$X \xrightarrow{\mathcal{U}_1} Z \xrightarrow{\mathcal{U}_2} Y,$$

where the diffeomorphism $\mathcal{U}_1 : \mathbb{R}^2/\Gamma_\tau \mapsto \mathbb{R}^2/\Gamma_\tau$ is completely defined by Lemma 3.6, and the diffeomorphism $\mathcal{U}_2 : \mathbb{R}^2/\Gamma_\tau \mapsto \mathbb{R}^2/\Gamma_1$ is unknown. We look for it in the form $Y = \mathcal{U}_2(Z)$ with

$$(3.27) \qquad y_1 = z_1 + d_2(Z), \quad y_2 = \tau(z_2 + e_2(z_2)),$$

where functions d_2 and e_2 will be specified below. Our first task is to make the formal change of variables in the left hand side of (3.13). We begin with the consideration of the second-order differential operator $\mathcal{J}^*(\mathfrak{a}^{-1}\mathcal{J}\cdot)$. It follows from the equality $\mathcal{J} = V \cdot \nabla_X$ that

$$-\mathcal{J}^*\big(\frac{1}{\mathfrak{a}}\mathcal{J}\check{u})\big) \circ \mathcal{U}_1^{-1} \circ \mathcal{U}_2^{-1} \equiv \frac{1}{J}\nabla_Y \cdot \Big\{\frac{J}{\mathfrak{a}}(\mathbb{B}^{-1}V \cdot \nabla_Y u)\mathbb{B}^{-1}V\Big\},$$

where

$$V = V(X(Y)), \quad \mathfrak{a} = \mathfrak{a}(X(Y)), \quad \mathbb{B}(Y) = \nabla_Y X(Y), J(Y) = \det \mathbb{B}(Y).$$

In particular, we have

$$\mathbb{B}^{-1}(Y(Z)) = \nabla_Z Y(Z) \left[\nabla_Z X(Z)\right]^{-1}.$$

On the other hand, Lemma 3.6 and formulae (3.27) imply

$$\left[\nabla_Z X(Z)\right]^{-1} V(X(Z)) = V_1(X(Z))\mathbf{e}_1, \quad \nabla_Z Y(Z)\,\mathbf{e}_1 = \big(1 + \partial_{z_1} d_2(Z)\big)\mathbf{e}_1,$$

where $\mathbf{e}_1 = (1,0)$. Thus we get

$$-\mathcal{J}^*\big(\frac{1}{\mathfrak{a}}\mathcal{J}\check{u})\big) \equiv \frac{1}{J}\partial_{y_1}\Big\{\frac{J}{\mathfrak{a}}(1+\partial_{z_1}d_2)^2 V_1^2 \partial_{y_1}u\Big\}.$$

From this and parametric representation (3.17) of the Dirichlet-Neumann operators we conclude that the left hand side of the desired identity (3.13) is equal to

$$\mathfrak{p}(Z(Y))\partial_{y_1}^2 u(Y) + \mathfrak{s}(Y))\partial_{y_1}u(Y) + \mathcal{G}_1 u(Y) + \mathcal{G}_0 u(Y) + \mathcal{G}_{-1}u(Y),$$

where

(3.28) $$\mathfrak{p}(Z) = \frac{1}{\mathfrak{a}(X(Z))}\big(1 + \partial_{z_1}d_2(Z)\big)^2 V_1(X(Z))^2,$$

$$\mathfrak{s}(Y) = \frac{1}{J(Y)}\partial_{y_1}\big\{J(Y)\mathfrak{p}(Z(Y))\big\}.$$

Note that (3.13) is a second order operator with respect to the variable y_1 and a first- order operator with respect to variable y_2. Its principal part is the pseudodifferential operator $\mathfrak{p}\partial_{y_1}^2 + \mathcal{G}_1$ which symbol reads

$$-\mathfrak{p}k_1^2 + J^{-1}\big(\text{adj}\,\mathbb{G}\,k \cdot k\big)^{1/2},$$

which we write in the form

$$-\mathfrak{p}k_1^2 + (\tau J)^{-1}\sqrt{g_{11}}\Big(\tau^2 k_2^2 - 2\tau^2 g_{11}^{-1}g_{12}k_1 k_2 + \tau^2 g_{11}^{-1}g_{22}k_1^2\Big)^{1/2}.$$

Hence operator (3.13) can be reduced to the operator with constant coefficients at principal derivatives if for some constant ν,

(3.29) $$\nu^{-1}\mathfrak{p}(Z(Y)) = (\tau J(Y))^{-1}\sqrt{g_{11}(Y)}.$$

This equality can be regarded as first order differential equation for functions $d_2(Z)$, $e_2(z_2)$ and a constant ν. It becomes clear if we write the right hand side as a function of the variable Z. To this end note that since $\partial_{y_1} z_2(Y) = \partial_{z_2} x_1(Z) = 0$,

$$g_{11} = \left|\nabla_Z X\, \partial_{y_1} Z\right|^2 + (\nabla_X \eta \cdot \nabla_Z X \cdot \partial_{y_1} Z)^2 =$$
$$[\partial_{y_1} z_1(Y)]^2 \left[1 + (\partial_{z_1}\mathcal{Z})^2 + (\partial_{x_1}\eta + \partial_{x_2}\eta \partial_{z_1}\mathcal{Z})^2\right],$$

which along with (3.25) gives

$$(3.30) \qquad g_{11} = \frac{V^2 + (V \cdot \nabla_X \eta)^2}{V_1^2 (1 + \partial_{z_1} d_2)^2}.$$

On the other hand, we have

$$J(Y(Z))^{-1} = \det \nabla_Z Y(Z)\bigl(\det \nabla_Z X(Z)\bigr)^{-1} = \tau(1+\partial_{z_1} d_2)(1+\partial_{z_2} e_2)(\partial_{z_2}\mathcal{Z})^{-1}.$$

Substituting these equalities into (3.29) we obtain the differential equation for d_2 and e_2,

$$(3.31) \qquad \partial_{z_1} d_2(Z) = \left[\nu \mathfrak{q}(Z)\bigl(1 + e_2'(z_2)\bigr)\right]^{1/2} - 1,$$

where $\mathfrak{q}$ is equal to
$$(3.32)$$
$$V_1(X(Z))^{-3}\Bigl(|V(X(Z))|^2 + \bigl(V(X(Z))\cdot \nabla_X \eta(X(Z))\bigr)^2\Bigr)^{1/2} \mathfrak{a}(X(Z))\partial_{z_2}\bigl(\mathcal{Z}(Z)\bigr)^{-1},$$

has symmetry (e, e), and is close to μ. Note that it is completely defined by equalities (3.1), (3.4), and Lemma 3.6. This equation can be solved as follows. We first note that it gives a unique d_2 with symmetry (o, e) provided that

$$(3.33) \qquad \nu^{-1/2}(1 + e_2'(z_2))^{-1/2} = \frac{1}{2\pi}\int_{-\pi}^{\pi} \mathfrak{q}^{1/2}(Z)\,dz_1.$$

This now determines a unique periodic odd function e_2 provided that

$$(3.34) \qquad \nu = 2\pi\tau \int_{-\pi/\tau}^{\pi/\tau}\left(\int_{-\pi}^{\pi}\mathfrak{q}^{1/2}(Z)\,dz_1\right)^{-2} dz_2,$$

which is just $1/\mu$ when $(\psi, \eta) = 0$. We observe that the invariance of $\mathfrak{q}(Z)$ under the shift $T_{\mathbf{v}_0}$, leads to the invariance of the right hand side of (3.33) by the change $z_2 \mapsto z_2 + \pi/\tau$. This means that e_2 is indeed $\pi/\tau-$ periodic. It follows from Appendix G (see Lemma G.3) of [**IPT**] and Lemma 3.6 that

$$\|\mathfrak{q} - \mu\|_{C^{m-4}} \leq c_m(\|U\|_4)\|U\|_m,$$
$$|\nu - 1/\mu| \leq c\|U\|_4,$$
$$\|d_2\|_{C^{m-4}} + \|e_2\|_{C^{m-4}} \leq c_m(\|U\|_4)\|U\|_m.$$

From this, the identities

$$d(X) = d_2(x_1, x_2 + d_1(X)), \quad e(X) = d_1(X) + e_2(x_2 + d_1(X)),$$

and from Appendix G (see Lemma G.1) of [**IPT**] we deduce estimates (3.15) for the functions d, e, and also estimate (3.16) for the Jacobian J. Hence the diffeomorphism $Y \to X$ is well defined and meets all requirements of assertion (iv) of Theorem 3.4. Next set

$$\kappa(Y) = (\tau J(Y))^{-1}\sqrt{g_{11}(Y)}, \quad b(Y) = -\tau g_{11}(Y)^{-1}g_{12}(Y),$$
$$1 + 2a = \tau^2 g_{11}(Y)^{-1}g_{12}(Y).$$

It follows from (3.15) that

$$(3.35) \quad \|g_{11} - 1\|_{C^{m-5}} + \|g_{22} - 1/\tau^2\|_{C^{m-5}} + \|g_{12}\|_{C^{m-5}} \leq c_m(\|U\|_5)\|U\|_m,$$

which leads to estimate (3.16) for κ and the following estimates for the functions a and b

$$(3.36) \quad \|a\|_{C^{m-5}} + \|b\|_{C^{m-5}} \leq c_m(\|U\|_5)\|U\|_m.$$

We are now in a position to define the operators $\mathfrak{A}$, $\mathfrak{B}$ and $\mathfrak{L}_{-1}$ defined by the left hand side of identity (3.13). It follows from (3.28) and (3.29), that its coefficients satisfies the inequalities.

$$\mathfrak{p} = \nu\kappa, \quad \mathfrak{p}^{-1}\mathfrak{s} = \partial_{y_1}(\ln \kappa J)$$

From this we conclude that

$$\left[\mathcal{G}_\eta \check{u} - \mathcal{J}^*\left(\frac{1}{\mathfrak{a}}\mathcal{J}\check{u}\right)\right] \circ (\mathcal{U}_2 \circ \mathcal{U}_1)^{-1} = \kappa\left[\mathfrak{L} + \widetilde{\mathcal{G}}_1 u + \nu \partial_{y_1}(\ln \kappa J)\mathfrak{D}_1 u + \mathfrak{B}u + \mathfrak{L}_{-1}u\right]$$

where

$$(3.37) \quad \mathfrak{B} = \frac{1}{\kappa}\mathcal{G}_0, \quad \mathfrak{L}_{-1} = \frac{1}{\kappa}\mathcal{G}_{-1}, \quad \widetilde{\mathcal{G}}_1 = \frac{1}{\kappa}\mathcal{G}_1 - (-\Delta)^{1/2}.$$

By construction, the operator $\widetilde{\mathcal{G}}_1$ has the following symbol

$$\left\{(1 + 2a(Y))k_1^2 + 2b(Y)\tau k_1 k_2 + \tau^2 k_2^2\right\}^{1/2} - \left\{k_1^2 + \tau^2 k_2^2\right\}^{1/2},$$

which implies that $\widetilde{\mathcal{G}}_1 = \mathfrak{A}_0 \mathfrak{D}_1$ where the zero order pseudodifferential operator $\mathfrak{A}_0$ has the symbol

$$\frac{-2i(a(Y)k_1 + \tau b(Y)k_2)}{\left\{(1 + 2a(Y))k_1^2 + 2b(Y)\tau k_1 k_2 + \tau^2 k_2^2\right\}^{1/2} + \left\{k_1^2 + \tau^2 k_2^2\right\}^{1/2}}.$$

which can be written in the standard form

$$(3.38) \quad A_0(Y, \xi) = \frac{-2i\big(a(Y)\xi_1 + b(Y)\xi_2\big)}{1 + \big(1 + 2a(Y))\xi_1^2 + 2b(Y)\xi_1\xi_2 + \xi_2^2\big)^{1/2}}.$$

Hence the basic identity (3.13) holds with the operators $\mathfrak{B}$, $\mathfrak{L}_{-1}$ and the operator $\mathfrak{A} = \mathfrak{A}_0 + \nu\partial_{y_1}(\ln g_{11})$ which yields (i). Estimates (ii) follow from formula (3.38), estimates (3.36), (3.35) (3.16) and Theorem 3.5.

Note that the functions κ, J, a are even and the function b is odd both in y_1 and y_2. Moreover, they are invariant under the shift $Y \to Y + (\pi, \pi/\tau)$. Hence the operator $\mathfrak{A}$ satisfies the symmetry conditions (iii). Formulae (3.37) along with Theorem 3.5 imply that the operators $\mathfrak{B}$ and $\mathfrak{L}_{-1}$ also satisfy the conditions (3.14), which completes the proof of Theorem 3.4.

3.3. Main orders of the diffeomorphism and coefficient ν

In using later the Nash-Moser theorem, we need to set

$$U = U_\varepsilon^{(N)} + \varepsilon^N W$$

where $U_\varepsilon^{(N)}$ is an approximate solution, up to order ε^N of the system (1.6), (1.7), and W is the unknown perturbation. At Theorem 2.3, for $\varepsilon_1 = \varepsilon_2 = \varepsilon/2$ (Diamond waves), we showed how to compute explicitly any order of approximation $U_\varepsilon^{(N)}$, with $\mu - \mu_c(\tau) = \varepsilon^2 \mu_1(\tau) + O(\varepsilon^4)$, $\mu_1(\tau) \neq 0$ for $\tau \neq \tau_c$. Then we need to know in particular the principal part of the coefficient ν in (3.34) and (3.9), which implies the knowledge of the diffeomorphisms $\mathcal{U}_1$ and $\mathcal{U}_2$. We show the following two Lemmas:

LEMMA 3.7. *For $N \geq 3$, we have*

$$(3.39) \qquad \nu(\varepsilon, \tau, W) = \mu_c^{-1} - \varepsilon^2 \nu_1(\tau) + O(\varepsilon^3),$$

where $\mu_c = (1 + \tau^2)^{-1/2}$, and

$$(3.40) \quad \begin{aligned} \nu_1(\tau) &= \frac{\mu_1(\tau)}{\mu_c^2} - \frac{3}{16\mu_c} + \frac{5}{4\mu_c^3} \\ &= \frac{1}{4\mu_c^5} - \frac{1}{2\mu_c^4} + \frac{1}{2\mu_c^3} + \frac{7}{8\mu_c^2} - \frac{1}{4\mu_c} - \frac{9}{16(2-\mu_c)}, \end{aligned}$$

which is positive for any $\tau > 0$, $(\nu_1(\tau) > 5/16\mu_c)$.

LEMMA 3.8. *For $N \geq 3$, the diffeomorphism (3.8) of the torus, which allows to obtain the form (3.13) for the linear operator $\mathcal{G}_\eta - \mathcal{J}^*(\frac{1}{\mathfrak{a}}\mathcal{J}(\cdot))$, and the principal part of $\tilde{\eta}(Y)$, are given by*

$$X = (x_1, x_2), \quad Y = (y_1, y_2), \quad X(Y) = (X_1(Y), X_2(Y)),$$

$$\begin{aligned} X_1(Y) &= y_1 + \frac{\varepsilon}{2} \sin y_1 \cos y_2 + \varepsilon^2 (\xi_{11} \sin 2y_1 + \xi_{12} \sin 2y_1 \cos 2y_2) + O(\varepsilon^3), \\ X_2(Y) &= \frac{y_2}{\tau} + \varepsilon \tau \cos y_1 \sin y_2 + \varepsilon^2 (\xi_{21} \sin 2y_2 + \xi_{22} \cos 2y_1 \sin 2y_2) + O(\varepsilon^3), \\ \tilde{\eta}(Y) &= \frac{-\varepsilon}{\mu_c} \cos y_1 \cos y_2 + \varepsilon^2 \{\eta_{00} + \eta_{01} \cos 2y_1 + \eta_{02} \cos 2y_2 (1 + \cos 2y_1)\} \\ &\quad + O(\varepsilon^3), \end{aligned}$$

where $\varepsilon > 0$ is defined by the main order $\varepsilon \xi_0$ of $U_\varepsilon^{(N)}$ and

$$\mu = \mu_c + \varepsilon^2 \mu_1(\tau) + O(\varepsilon^4), \quad \mu_c = (1 + \tau^2)^{-1/2},$$

and where
$$\xi_{11} = -\frac{1}{4}\left\{\frac{3}{4(2-\mu_c)} - \frac{3}{16} - \frac{5}{4\mu_c} + \frac{3}{4\mu_c^2} + \frac{1}{\mu_c^3} - \frac{1}{\mu_c^4}\right\},$$
$$\xi_{12} = -\frac{1}{64}, \quad \xi_{21} = -\frac{\tau}{8} + \frac{17}{32\tau}, \quad \xi_{22} = -\frac{\tau}{8},$$
$$\eta_{00} = \frac{1}{4\mu_c^3} - \frac{1}{8\mu_c}, \quad \eta_{02} = \frac{1}{8\mu_c},$$
$$\eta_{01} = -\frac{1}{4\mu_c} - \frac{1}{4\mu_c^2} + \frac{1}{4\mu_c^3} + \frac{3}{8(2-\mu_c)}.$$

Proof: see Appendix D.

CHAPTER 4

Small Divisors. Estimate of $\mathfrak{L}-$ Resolvent

It follows from Theorem 3.4 that the pseudodifferential operator $\mathfrak{L} = \mathcal{G}^{(0)} + \nu\partial_{y_1}^2$ with the symbol

(4.1) $$L(k) = -\nu k_1^2 + (k_1^2 + \tau^2 k_2^2)^{1/2}, \quad k \in \mathbb{Z}^2.$$

forms the principal part of the linear problem. In this chapter we describe the structure of the spectrum of $\mathfrak{L}$ and investigate in details the dependence of its resolvent on parameters ν and τ. The difficulty here, is that we need to play on two parameters, taking into account that the small divisor problem already occurs at criticality (origin of the half line). The first result in this direction is the following theorem which constitutes generic properties of $\mathfrak{L}$.

THEOREM 4.1. *For every positive ν and τ, the operator $\mathfrak{L}$ is selfadjoint in $H^0(\mathbb{R}^2/\Gamma)$ and has the natural domain of definition*

$$D(\mathfrak{L}) = \{u \in H^0(\mathbb{R}^2/\Gamma) : \sum_{k \in \mathbb{Z}^2} L(k)^2 |\widehat{u}(k)|^2 < \infty\}.$$

The spectrum of $\mathfrak{L}$ coincides with the closure of the discrete spectrum which consists of all numbers $\lambda_k = L(k)$, $k \in \mathbb{N}^2$; the corresponding eigenfunctions are defined by

$$(2\pi)^{-1}\exp(\pm ik \cdot Y), \quad (2\pi)^{-1}\exp(\pm k^* \cdot Y) \text{ where } k^* = (-k_1, k_2).$$

For each real $\varkappa \neq \lambda_k$, $k \in \mathbb{Z}^2$, the resolvent $(\mathfrak{L} - \varkappa)^{-1}$ is a selfadjoint unbounded operator defined in terms of the Fourier transform by the formula

(4.2) $$\widehat{(\mathfrak{L} - \varkappa)^{-1}u}(k) = (L(k) - \varkappa)^{-1}\widehat{u}(k), \quad k \in \mathbb{Z}^2.$$

PROOF. The operator $\mathfrak{L}$ is unitary equivalent to a multiplication operator in the Hilbert space l_2, which can be represented by the infinite diagonal matrix with diagonal elements $L(k)$, $k \in \mathbb{Z}^2$. It remains to note that for such a matrix the spectrum coincides with the closure of diagonal, and the discrete spectrum coincides with the diagonal. □

Note that zero is nontrivial eigenvalue of $\mathfrak{L}$ if and only if the dispersion equation $L(k) = 0$ has a nontrivial solution $k = k_0$. The number of such solutions depends on the arithmetic nature of parameters ν, τ. We restrict our attention to the simplest case when $k_0 = (1,1)$. It is easy to see that in this case the point $(\nu, \tau) = (\nu_0, \tau)$ belongs to the positive branch of the hyperbola $\nu_0^2 - \tau^2 = 1$.

Our aim is to investigate in details the dependence of the $\mathfrak{L}$-resolvent on parameter ν. With further applications to the Nash-Moser theory in mind we take the perturbed values of parameter ν, and a spectral parameter $\varkappa$ in the form

(4.3) $$\nu_j(\varepsilon) = \nu_0 - \varepsilon^2 \nu_1 + \varepsilon^3 \tilde{\nu}_j(\varepsilon), \quad \varkappa_j(\varepsilon) = \varepsilon^2 \tilde{\varkappa}_j(\varepsilon),$$

where $\nu_0 = \nu_0(\tau)$ and $\nu_1 = \nu_1(\tau)$. Here functions $\tilde{\nu}_j$ and $\tilde{\varkappa}_j$, $j \geq 1$, are defined on the segment $[0, r_0]$ and satisfy the inequalities

(4.4)
$$|\tilde{\nu}_j(\varepsilon)| + |\tilde{\varkappa}_j(\varepsilon)| \leq R,$$
$$|\tilde{\nu}_j(\varepsilon') - \tilde{\nu}_j(\varepsilon'')| + |\tilde{\varkappa}_j(\varepsilon') - \tilde{\varkappa}_j(\varepsilon'')| \leq R|\varepsilon' - \varepsilon''|,$$
$$|\tilde{\nu}_{j+1} - \tilde{\nu}_j| + |\tilde{\varkappa}_{j+1} - \tilde{\varkappa}| \leq R(2^{-j}).$$

Denoting by $\mathfrak{L}_0$ the operator $\mathfrak{L}$ for $\nu = \nu_0(\tau)$, the next theorem establishes the basic estimates for a resolvent $(\mathfrak{L} - \varkappa)^{-1}$ on the orthogonal complement to ker $\mathfrak{L}_0$, which are stable with respect to perturbations of parameters from a suitable Cantor set.

THEOREM 4.2. (a) *For each $\alpha \in (0, 1]$, there is a set of full measure $\mathfrak{N}_\alpha \subset (1, \infty)$ so that whenever $\nu_0 \in \mathfrak{N}_\alpha$ and $\nu = \nu_0(\tau) = (1+\tau^2)^{1/2}$, zero is a non-trivial eigenvalue of $\mathfrak{L}_0$ and*

$$\ker \mathfrak{L}_0 = \text{span} \{1, \exp(\pm i k_0 \cdot Y), \exp(\pm k_0^* \cdot Y)\},$$

(4.5) $\quad \|\mathfrak{L}_0^{-1} u\|_{s-(1+\alpha)/2} \leq c \|u\|_s$, *when $u \in (\ker \mathfrak{L}_0)^\perp \cap H^s(\mathbb{R}^2/\Gamma)$.*

(b) *Suppose that $\nu_0 \in \mathfrak{N}_\alpha$ with $\alpha \in (0, 1/78)$, then there is a set $\mathcal{E} \subset [0, r_0)$ so that*

(4.6) $$\frac{2}{r^2} \int_{[0,r] \cap \mathcal{E}} \varepsilon \, d\varepsilon \to 1 \text{ as } r \to 0,$$

and for all $\varepsilon \in \mathcal{E}$, $j, s \geq 1$ and $\nu = \nu_j(\varepsilon)$, $\varkappa = \varkappa_j(\varepsilon)$,

(4.7) $\quad \|(\mathfrak{L} - \varkappa)^{-1} u\|_{s-1} \leq c \|u\|_s$ *when $u \in (\ker \mathfrak{L}_0)^\perp \cap H^s(\mathbb{R}^2/\Gamma)$,*

where the positive constant c depends on α, ν_0 and R only.

REMARK 4.3. *The fact proved at Lemma 3.7 that $\nu_1(\tau) > 0$, allows the freedom to take τ as a function of ε as $\tau = \tau_0 + \varepsilon^2 \tau_1$. Then, for $|\tau_1|$ small enough, we have*

$$\omega_1 := \omega_0(\nu_0^{-1} \nu_1 - \tau_0^{-1} \tau_1) \neq 0$$

and Theorem 4.2 still holds true.

It follows from formula (4.2) and the Parseval identity that Theorem 4.2 will be proved once we prove the following

THEOREM 4.4. (a) *For any $\alpha \in (0, 1]$ there is a set of full measure $\mathfrak{N}_\alpha \subset (1, \infty)$ so that for all $\nu_0 \in \mathfrak{N}_\alpha$ and $\nu = \nu_0(\tau)$,*

(4.8) $\quad |L(k)| \geq c|k|^{-(1+\alpha)/2}$ *when $k \neq 0, \pm k_0, \pm k_0^*$.*

(b) *Suppose that $\nu_0 \in \mathfrak{N}_\alpha$ with $\alpha \in (0, 1/78)$, then, there is a set $\mathcal{E} \subset [0, r_0)$ satisfying (4.6) so that for all $\varepsilon \in \mathcal{E}$, $j \geq 1$ and $\nu = \nu_j(\varepsilon)$, $\varkappa = \varkappa_j(\varepsilon)$,*

(4.9) $$\left|(L(k) - \varkappa)^{-1}\right| \geq c|k|^{-1} \text{ for all } k \neq 0, \pm k_0, \pm k_0^*,$$

where the positive constant c depends on ν_0, α and R only.

The proof naturally falls into four steps.

First step. We begin with proving two auxiliary lemmas which establish a connection between the symbol L and the Diophantine function defined by the formula

(4.10) $$(m, n) \to \omega - \frac{m}{n^2} - \frac{C}{n^2}, \quad (m, n) \in \mathbf{N}^2,$$

where constants ω, C are connected with parameters ν, τ and $\varkappa$ by the relations

$$\omega = \nu/\tau, \quad C = 1/(2\nu\tau) - \varkappa/\tau.$$

LEMMA 4.5. *Let for some $\varrho \geq 2$ and $\alpha \in [0, 1]$, parameters ν, τ and a vector $k \in \mathbb{Z}^2$ satisfy the inequalities*

(4.11) $$0 < \varrho^{-1} \leq \nu, \tau < \varrho, \quad |\varkappa| < \varrho, \quad |k_1| \geq 2\varrho^2,$$

(4.12) $$\left|\omega k_1^2 - |k_2| - C\right| \geq 5(\varrho^2 + \varrho^{10})|k_1|^{-1-\alpha},$$

Then $|L(k) - \varkappa| \geq |k|^{-(1+\alpha)/2}$.

PROOF. Since $L(k)$ is even in k, it suffices to prove the lemma for

$$k = (n, m) \text{ with integers } n > 2\varrho^2, \quad m > 0.$$

We begin with the observation that for $n^2 \geq 2\varrho^2 m$,

$$-L(k) + \varkappa \geq \varrho^{-1} n^2 - \sqrt{\varrho^2 m^2 + n^2} - \varrho \geq$$
$$4\varrho^4 \left(\frac{1}{\varrho} - \frac{1}{2}\left(\frac{1}{\varrho^2} + \frac{1}{\varrho^4}\right)^{1/2}\right) - \varrho \geq \frac{4}{3}\varrho^3 - \varrho > 1.$$

Hence it suffices to prove the lemma for

(4.13) $$n^2 \leq 2\varrho^2 m.$$

Since for all non-negative σ, $0 \leq 1 + \sigma/2 - (1 + \sigma)^{1/2} \leq \sigma^2/4$, we have

$$-\tau^{-1} L(k) = \omega n^2 - m - (2\tau^2 m)^{-1} n^2 + (4\tau^4 m^3)^{-1} n^4 \, o_1, \quad o_1 \in [0, 1],$$

which along with the identity

$$(2\tau^2 m)^{-1} n^2 = (2\nu\tau)^{-1} + (2\nu\tau m)^{-1}(\omega n^2 - m)$$

yields

(4.14) $$-\tau^{-1} e(k)\bigl(L(k) - \varkappa\bigr) = \omega n^2 - m - C + o_2.$$

Here

$$o_2 = C(1 - e(k)) + e(k)(4\tau^4 m^3)^{-1} n^4 o_1, \quad e(k) = \bigl(1 - (2\tau\nu m)^{-1}\bigr)^{-1}.$$

Since
$$(2\nu\tau m)^{-1} \leq 2^{-1}\varrho^2 m^{-1} \leq \varrho^4 n^{-2} \leq 2^{-1},$$
we have
$$1 \leq e(k) \leq 2, \quad e(k) - 1 \leq 2\varrho^4 n^{-2},$$
which along with the inequality $|C| \leq 3\varrho^2/2$ leads to the estimate
$$|o_2| \leq \left(3\varrho^6 + \frac{\varrho^4}{2}\frac{n^6}{m^3}\right)\frac{1}{n^2} \leq 5\varrho^{10} n^{-2}.$$
From this, (4.12) and (4.14) we conclude that
$$2\varrho|L(k) - \varkappa| \geq |\omega n^2 - m - C| - 2\varrho^{10} n^{-2} \geq$$
$$5(\varrho^2 + \varrho^{10})n^{-1-\alpha} - 5\varrho^{10}n^{-2} \geq 5\varrho^2 n^{-1-\alpha}.$$
It remains to note that due (4.13) the right hand side is larger than $|k|^{-(1+\alpha)/2}$ and the lemma follows. $\square$

LEMMA 4.6. *Suppose that parameters ν_0, τ_0 and $\nu, \tau, \varkappa$ satisfy the following conditions.*

(i) *There is $\varrho \geq 2$ so that*
$$0 < \varrho^{-1} \leq \nu, \tau, \nu_0 < \varrho, \quad |\varkappa| < \varrho.$$

(ii) *The dispersion equation*
$$\nu_0 k_1^2 - \sqrt{k_1^2 + \tau^2 k_2^2} = 0 \tag{4.15}$$
has the unique positive solution $k = k_0$.

(iii) *There are positive N and q so that*
$$N \geq 2\varrho^2, \quad q \geq 5(\varrho^2 + \varrho^{10}),$$
$$|\omega n^2 - m - C| \geq q n^{-1-\alpha} \text{ for all integers } n \geq N, m \geq 0$$

Then there are positive δ and γ_0, depending on N, q and ν_0 only, so that
$$|L(k) - \varkappa| \geq \gamma_0 |k|^{-(1+\alpha)/2} \text{ for all } k \neq 0, \pm k_0, \pm k_0^* \tag{4.16}$$
when
$$|\nu - \nu_0| + |\varkappa| \leq \delta. \tag{4.17}$$

PROOF. It follows from (iii) that ν, τ and $\varkappa$ meet al requirements of Lemma 4.5 which implies that $|L(k) - \varkappa| \geq |k|^{-(1+\alpha)/2}$ for all $|k_1| \geq N$. On the other hand, since
$$\lim_{|k_2| \to \infty} |L(k) - \varkappa| \geq \lim_{|k_2| \to \infty} \sqrt{\varrho^{-2} k_2^2 + k_1^2} - \varrho k_1^2 - \varrho = \infty,$$
there is N^* depending on ϱ and N only so that $|L(k) - \varkappa| \geq |k|^{-(1+\alpha)/2}$ for all $|k| \geq N^*$. Since the number of wave vectors k in the circle $|k| \leq N$ is finite, the existence δ and γ_0 follows from continuity of $L(k)$ as a function of parameters ν, τ. $\square$

Second step. Next we prove that condition (iii) from the previous lemma is the generic property of function (4.10) which leads to assertion (a) of Theorem 4.4. In order to formulate the results let us introduce the important function $d: \mathbb{N}^2 \mapsto \mathbb{R}$ defined by the formula

$$(4.18) \qquad d(m,n) = \omega_0 - \frac{m}{n^2} - \frac{C_0}{n^2}, \text{ where } \omega_0 = \nu_0/\tau, C_0 = (2\nu_0\tau)^{-1}.$$

LEMMA 4.7. *For each $\alpha \in (0,1]$ and $q > 0$ there is a set of a full measure $\mathfrak{M}_\alpha$ in $(1,\infty)$ so that for all $\nu_0 \in \mathfrak{M}_\alpha$, there exists $N > 0$, depending on ν_0, α and q only, such that*

$$(4.19) \qquad |d(m,n)| \geq qn^{-3-\alpha} \text{ when } n \geq N, m \geq 0.$$

PROOF. Noting that $C_0 = (\omega_0 - \omega_0^{-1})/2$, we rewrite inequality (4.19) in the equivalent form

$$|d_0(\omega_0,m,n)| \geq qn^{-3-\alpha} \text{ where } d_0(\omega_0,m,n) = \omega_0 - \frac{1}{n^2}\left(m - 2^{-1}\omega_0 + 2^{-1}\omega_0^{-1}\right).$$

Without loss of generality we can assume also that $\omega_0 \in [1,a]$, where a is an arbitrary positive number. The set of points ω_0 for which $|d_0(\omega_0,m,n)| \leq qn^{-3-\alpha}$ can be covered by the system of the intervals $\iota(m,n)$ labelled by integers m and n. It is easy to see that their extremities $\omega_0^\pm(m,n)$ satisfy $\left(n^2 + \frac{1}{2}\right)\omega_0^\pm(m,n) \geq m - \frac{q}{n^{1+\alpha}}$. On the other hand, since

$$\partial_\omega d_0(\omega_0,m,n) = 1 + (2n^2)^{-1} + (2\omega_0^2 n^2)^{-1} \in [1,2],$$

the length of each intervals is less than $2qn^{-3-\alpha}$. Hence for fixed n, the number of intervals $\iota(m,n)$ having nonempty intersections with a segment $[1,a]$ is less than $a(n^2 + 1/2) + q/n^{1+\alpha}$. From this we conclude that

$$\sum_{n \geq N} \sum_{m: \iota(m,n) \cap [1,a] \neq \emptyset} \text{meas } \iota(m,n) \leq 2q(\frac{3}{2}a + q) \sum_{n \geq N} n^{-1-\alpha} \leq cq(a+q)N^{-\alpha}.$$

Hence

$$\text{meas } \{\omega_0 \in [1,a] : |d_0(\omega,m,n)| \geq qn^{-3-\alpha} \text{ for all } m \geq 0, n \geq N\} \geq$$
$a - cq(a+q)N^{-\alpha} \to a$ as $N \to \infty$.

It remains to note that the mapping $\nu_0 \to \omega_0$ takes diffeomorphically the interval $[1,\infty)$ onto itself and the lemma follows. $\square$

Next lemma shows that almost each point of $\mathfrak{M}_\alpha$ satisfies the following

Condition M. *There are positive constant c_0, c_1 such that for all integers $n > 0$, $m \geq 0$,*

$$(4.20) \qquad |d(m,n)| \geq c_0 n^{-3-\alpha},$$

$$(4.21) \qquad \left|e^{2\pi i \omega_0 n} - 1\right|^{-1} \leq c_1 n^2$$

LEMMA 4.8. *For each $\alpha \in (0,1)$ there is a set $\mathfrak{N}_\alpha \subset \mathfrak{M}_\alpha$ such that meas $(\mathfrak{M}_\alpha \setminus \mathfrak{N}_\alpha) = 0$ and Condition M holds for each $\nu_0 \in \mathfrak{N}_\alpha$.*

PROOF. We begin with the observation that inequality (4.21) holds for almost every ω_0 with a constant c_1 depending on ω_0 only. Hence it suffices to show that inequality (4.20) holds true for each $\nu_0 \in \mathfrak{M}_\alpha$. Fix $q = 1$, $\nu \in \mathfrak{M}_\alpha$ and note that by lemma (4.7), $|d(m,n)| \geq n^{-3-\alpha}$ for all $n \geq N$. Since $\lim_{m \to \infty} d(m,n) = -\infty$, we can choose N such that this inequality is fulfilled for all (m,n) outside of the simplex $n > 0, m \geq 0, n + m \leq N$. It remains to note that this simplex contains only finite number of integer points and $d(m,n) \neq 0$ for irrational ω_0 different from the sum of a rational and the square root of a rational. □

Third step. We intend now to study the robustness of estimate (4.20) when one adds a small perturbation to $d(m,n)$. In order to formulate the corresponding result we introduce some notations. For each $\varepsilon \in [0, r_0]$ and $j \geq 1$ set
$$\tilde{\omega}_j(\varepsilon) = \nu_j(\varepsilon)/\tau, \quad \tilde{C}_j(\varepsilon) = \bigl(2\nu_j(\varepsilon)\tau\bigr)^{-1} - \varkappa_j(\varepsilon)\tau^{-1}$$
It follows from (4.3) that they have the representation
$$\tilde{\omega}_j(\varepsilon) = \omega_0 - \varepsilon^2 \omega_1 + \varepsilon^3 \tilde{\Omega}_j(\varepsilon), \tilde{C}_j(\varepsilon) = C_0 - \varepsilon^2 \tilde{\varphi}_j(\varepsilon) \quad \varepsilon \in [0, r_0],$$
in which $\omega_1 = \tau^{-1}\nu_1$. Our task is to obtain the estimates for the function $\tilde{D}_j : [0, r_0] \times \mathbb{N}^2 \to \mathbb{R}$ defined by

(4.22) $$\tilde{D}(\varepsilon, m, n,) = \tilde{\omega}_j(\varepsilon) - \frac{m}{n^2} - \frac{C_0}{n^2} + \varepsilon^2 \frac{\tilde{\varphi}_j(\varepsilon)}{n^2}.$$

For technical reason it is convenient to formulate the problem in terms of a new small parameter λ. Set
$$\lambda = \varepsilon^2, \quad \varphi_j(\lambda) = \tilde{\varphi}_j(\sqrt{\lambda}), \quad \Omega_j(\lambda) = \tilde{\Omega}_j(\sqrt{\lambda}),$$
(4.23) $$\omega_j(\lambda) = \omega_0 - \lambda\omega_1 + \lambda^{3/2}\Omega_j(\lambda), \quad \lambda \in [0, \rho_0],$$
where $\rho_0 = r_0^2$. It follows from (4.4) that for suitable choice R and r_0,
(4.24)
$$|\Omega_j|+|\varphi_j| \leq R, \quad |\Omega_{j+1}-\Omega_j|+|\varphi_{j+1}-\varphi_j| \leq R(2^{-j}), \quad |\Omega'_j(\lambda)|+|\varphi'_j(\lambda)| \leq \frac{R}{2\sqrt{\lambda}}.$$

Set

(4.25) $$D_j(\lambda, m, n,) := \tilde{D}_j(\sqrt{\lambda}, m, n) = \omega_j(\lambda) - \frac{m}{n^2} - \frac{C_0}{n^2} + \lambda\frac{\varphi_j(\lambda)}{n^2}.$$

DEFINITION 4.9. *For positive N, q, r denote by $\mathcal{H}_j(N, q, r)$ the set defined by*
(4.26)
$$\mathcal{H}_j(N,q,r) = \left\{ \lambda \in (0,r) : |D_j(\lambda, m, n)| \geq qn^{-4} \text{ for all integers } n \geq N, \; m \geq 0 \right\}.$$

THEOREM 4.10. *Assume that* $\nu_0 \in \mathfrak{N}_\alpha$ *with* $\alpha \in (0, 1/78)$. *Then, for each* $q > 0$, *there is* $N > 0$ *such that*

(4.27) $$\frac{1}{r} \text{ meas } \bigcap_{j \geq 1} \{\mathcal{H}_j(N, q, r)\} \to 1 \text{ as } r \to 0,$$

and there exists c^* *such that*

(4.28) $$\frac{1}{r} \text{ meas } \bigcap_{j \geq 1} \{\mathcal{H}_j(N, q, r)\} \geq 1 - c^* r^\varpi \text{ for } 0 < r < r^*,$$

where $0 < \varpi < \min\{\alpha/(3-\alpha), (78^{-1} - \alpha)/(3+\alpha)\}$.

PROOF. Section 4.1 is devoted to the proof. □

Fourth step. We are now in a position to complete the proof of Theorem 4.4. First note that by Lemmas 4.7 and 4.8 for each $\nu_0 \in \mathfrak{N}_\alpha$, parameters $\nu = \nu_0(\tau)$ and $\varkappa = 0$ meet all requirements of Lemma 4.6 with $\delta = 0$. Applying this lemma we obtain (4.8). It remains to note that for almost all ν_0 the dispersion equation has the only positive solution k_0 and assertion (a) follows. In order to prove (b) choose an arbitrary $\nu_0 \in \mathfrak{N}_\alpha$, $q > 0$ and set

$$\mathcal{E} = \{\varepsilon > 0 : \varepsilon^2 \in \bigcap_{j \geq 1} \mathcal{H}_j(N, q, r)\},$$

where N is given by Theorem 4.10. It follows from this theorem that the set $\mathcal{E}$ satisfies density condition (4.6). On the other hand, (4.26) implies that for $\varepsilon \in \mathcal{E}$, parameters $\nu = \nu_j(\varepsilon)$, $\varkappa = \varkappa_j(\varepsilon)$ meet all requirements of Lemma 4.6. Since

$$|\nu_j(\varepsilon) - \nu_0| + |\varkappa_j(\varepsilon)| \to 0 \text{ as } \varepsilon \to 0$$

uniformly with respect to j, there exists $\varepsilon_2 > 0$ depending on ν_0 and R only such that for all $\varepsilon \in \mathcal{E} \cap (0, \varepsilon_2)$, parameters $\nu = \nu_j(\varepsilon)$ and $\varkappa = \varkappa_j(\varepsilon)$ satisfy inequality (4.17) which yields (4.16) (where $\alpha = 1$), and the theorem follows.

4.1. Proof of Theorem 4.10

Our approach is based on standard methods of metric theory of Diophantine approximations and the Weil Theorem on the uniform distribution of a sequence $\{\omega_0 n^2\}$. Without loss of generality we can assume that $\omega_1 > 0$, and $\rho_0 = r_0^2$ satisfies the inequality

(4.29) $$\rho_0 < \frac{\omega_0}{4}(\omega_1 + R)^{-1} \Rightarrow \omega(\lambda) \geq \frac{3\omega_0}{4} \text{ for } \lambda \leq \rho_0.$$

It follows from (4.24) that the sequences Ω_j and φ_j converge uniformly on $[0, \rho_0]$ to functions Ω_∞ and φ_∞ such that
(4.30)
$$|\Omega_\infty| + |\varphi_\infty| \leq R, |\Omega_\infty - \Omega_j| + |\varphi_\infty - \varphi_j| \leq 2^{1-j} R, |\Omega'_\infty(\lambda)| + |\varphi'_\infty(\lambda)| \leq 2^{-1} \lambda^{-1/2} R.$$

Our task is to estimate the measure of intersection of the sets $\mathcal{H}_j$. We begin with the investigation of their structure.

Structure of a set $\mathcal{H}_j(N,q,r)$. The main result of this paragraph is the following covering lemma. Fix an arbitrary positive q and set
(4.31)
$$r_2 = \min\left\{\frac{\rho_0}{2}, \frac{\omega_1^2}{128R^2}\right\}, \quad N_3(q) = \max\left\{\left(\frac{16R}{\omega_1}\right)^{1/2}, \left(\frac{2q}{c_0}\right)^{1/(1-\alpha)}, \left(\frac{4q}{\omega_1 r_2}\right)^{1/4}\right\},$$
where c_0 is the constant from Condition M.

LEMMA 4.11. *For $N > N_3(q)$, $r < r_2$ and $1 \leq j \leq \infty$, the set $[0,r] \setminus \mathcal{H}_j(N,q,r)$ is covered by the system of the intervals*
$$I_j(m,n) = (\lambda_j^-(m,n), \lambda_j^+(m,n)), \quad n \geq N \quad I_j(m,n) \cap [0,r] \neq \emptyset,$$
such that

(i) $\lambda_j^\pm(m,n)$ are solutions of the equations

(4.32) $$\lambda_j^\pm\left(\omega_1 - \frac{\varphi_j(\lambda_j^\pm)}{n^2}\right) - (\lambda_j^\pm)^{3/2}\Omega(\lambda_j^\pm) = d(m,n) \pm \frac{q}{n^4}$$

with $d(m,n) > 0$. They satisfy the inequalities

(4.33) $$0 < \lambda_j^-(m,n) < \lambda_j^+(m,n) < 2r_2,$$

(4.34) $$\frac{2}{5\omega_1}d(m,n) \leq \lambda_j^\pm \leq \frac{2}{\omega_1}d(m,n)$$

(4.35) $$\frac{4q}{3\omega_1}\frac{1}{n^4} \leq \lambda_j^+(m,n) - \lambda_j^-(m,n) \leq \frac{4q}{\omega_1}\frac{1}{n^4}.$$

(ii) For a fixed $n > N$, the left extremities $\lambda_-^j(m,n)$ strongly decreases in m,

(4.36) $$\lambda_j^-(M_{j,n}(r), n) \leq \lambda_j^-(M_{j,n}(r)-1, n) \leq \ldots \leq \lambda^-(m_{j,n}(r), n),$$

(4.37) $$\lambda_j^-(k-1,n) - \lambda_j^-(k,n) \geq \frac{2}{3\omega_1 n^2},$$

where
$$m_{j,n}(r) = \min\{m > 0 : I_j(m,n) \cap [0,r] \neq \emptyset\},$$
$$M_{j,n}(r) = \max\{m > 0 : I_j(m,n) \cap [0,r] \neq \emptyset\}.$$

(iii) For each such interval with $I_j(m,n) \cap [0,r] \neq \emptyset$,

(4.38) $$d(m,n) \leq \frac{5}{2}\omega_1 r, \quad n \geq \left(\frac{2c_0}{5\omega_1}\right)^{1/(3+\alpha)} r^{-1/(3+\alpha)}, \lambda_j^+(m,n) \leq 2r.$$

(iv) If intervals $I_j(m,n)$ and I_∞ have nonempty intersections with $(0,r]$, then

(4.39) $$|\lambda_j^\pm(m,n) - \lambda_\infty^\pm(m,n)| \leq 2^{-j-4}\omega_1.$$

PROOF. By abuse of notations, we suppress the index j. Proof of (i).
$$-\partial_\lambda D(\lambda,m,n) = \omega_1 - \frac{\varphi + \lambda\varphi'}{n^2} - \frac{3}{2}\lambda^{1/2}\Omega(\lambda) - \lambda^{3/2}\Omega'(\lambda),$$

it follows from (4.31) and (4.24) that for $\lambda \leq 2r_2$
$$\omega_1 - \frac{2R}{n^2} - 2R\lambda^{1/2} \leq -\partial_\lambda D(\lambda, m, n) \leq \omega_1 + \frac{2R}{n^2} + 2R\lambda^{1/2},$$
which along with (4.31) implies the inequalities

(4.40) $\quad \dfrac{\omega_1}{2} \leq -\partial_\lambda D(\lambda, m, n) \leq \dfrac{3\omega_1}{2} \quad \text{for } \lambda \in [0, 2r_2], \quad n > N_3.$

Hence the function $-D(\lambda, m, n)$ is strongly monotone on the interval $(0, 2r_2)$. Therefore for $r < r_2, N \geq N_3$ the set $(0, r) \setminus \mathcal{H}(N, q, r)$ can be covered by the system of the intervals
$$J(m, n) = \big(\beta(m, n), \gamma(m, n)\big), \quad 0 \leq \beta(m, n) < \gamma(m, n) \leq r,$$
such that

(4.41) $\quad -D(\beta(m, n), m, n) = -\dfrac{q}{n^4} \quad \text{if } \beta(m, n) > 0,$

(4.42) $\quad -D(\gamma(m, n), m, n) = +\dfrac{q}{n^4} \quad \text{if } \gamma(m, n) < r,$

(4.43) $\quad -\dfrac{q}{n^4} \leq -D(\beta(m, n), m, n) < \dfrac{q}{n^4} \quad \text{if } \beta(m, n) = 0$

(4.44) $\quad -\dfrac{q}{n^4} < -D(\gamma(m, n), m, n) \leq \dfrac{q}{n^4} \quad \text{if } \gamma(m, n) = r,$

Note that, by condition M,
$$\frac{q}{n^4} = \frac{q}{c_0 n^{1-\alpha}} \frac{c_0}{n^{3+\alpha}} \leq \frac{q}{c_0 n^{1-\alpha}} |d(m, n)|,$$
which along with (4.31) yields the inequalities
(4.45)
$$\frac{1}{2}|d(m,n)| \leq |d(m,n) \pm \frac{q}{n^4}| \leq \frac{3}{2}|d(m,n)|, \quad |d(m,n)| \geq 2\frac{q}{n^4} \quad \text{for } n \geq N_3.$$

From this and the equality $D(0, m, n) = d(m, n)$ we conclude that case (4.43) is impossible and $\beta(m, n) > 0$ for all intervals $J(m, n)$. Let us consider case (4.44). Since the function $-D(\lambda, m, n)$ increases in λ on the segment $[0, 2r_2]$, there is a maximal $\gamma*$ in $(0, 2r_2]$ such that
$$\begin{aligned}(\beta(m,n), r) &\subset (\beta(m,n), \gamma^*) \subset (\beta(m,n), 2r_2), \\ |D(\lambda, m, n)| &\leq qn^{-4} \quad \text{in } (\beta(m,n), \gamma^*).\end{aligned}$$

Let us show that $\gamma^* < 2r_2$. If the assertion is false, then $r_2, 2r_2 \in (\beta(m, n), \gamma*]$ which yields
$$-qn^{-4} \leq -D(r_2, m, n) \leq -D(2r_2, m, n) \leq qn^{-4}.$$
Thus we get
$$r_2 \min_{[r_2, 2r_2]}\{-\partial_\lambda D(\lambda, m, n)\} \leq D(r_2, m, n) - D(2r_2, m, n) \leq 2qn^{-4} \leq 2qN_3^{-4}.$$

From this and (4.40) we obtain the inequality
$$\omega_1 \frac{r_2}{2} \leq 2qN_3^{-4},$$

which contradicts (4.31), and the assertion follows. In particular, we have $-D(\gamma^*, m, n) = qn^{-4}$. Since the equations $-D(\lambda_\pm, m, n) = \pm qn^{-4}$ are equivalent to (4.32), in the cases (4.41), (4.42) we have in the cases (4.41), (4.42) that

$$\beta(m,n) = \lambda^-(m,n), \quad \gamma(m,n) = \lambda^+(m,n), \quad J(m,n) = I(m,n),$$

in the case (4.44) we have

$$\beta(m,n) = \lambda^-(m,n), \quad \gamma^* = \lambda^+(m,n), \quad J(m,n) \subset I(m,n).$$

Hence the set $[0,r] \setminus \mathcal{H}(N,q,r)$ is covered by the intervals $I(m,n)$ with the ends satisfying (4.32), (4.33). It follows from (4.31) and (4.24) that for $n > N_3(q)$, $\lambda < r_2$,

(4.46) $$\frac{3\omega_1}{4}\lambda \leq \lambda\left(\omega_1 - \frac{\varphi}{n^2}\right) - \lambda^{3/2}\Omega(\lambda) \leq \frac{5}{4}\omega_1 \lambda.$$

In particular, the left side of equation (4.32) is positive, which along with (4.45) yields positivity of $d(m,n)$ in the right side of equations (4.32). Combining (4.32), (4.46), (4.45) gives (4.34). Next equation (4.32) implies

$$D(\lambda^-(m,n), m, n) - D(\lambda^+(m,n), m, n) = 2qn^{-4}.$$

From this and (4.40) we conclude that

$$\omega_1(\lambda^+(m,n) - \lambda^-(m,n)) \leq 4qn^{-4},$$
$$\omega_1(\lambda^+(m,n) - \lambda^-(m,n)) \geq \frac{4}{3}qn^{-4},$$

which yields

$$\frac{4q}{3\omega_1}\frac{1}{n^4} \leq \lambda^+(m,n) - \lambda_-(m,n) \leq \frac{4q}{\omega_1}\frac{1}{n^4}$$

and the assertion follows.

Let us turn to the proof of (ii). By Lemma 4.11(i), $I(m,n) \cap [0,r] \neq \emptyset$ if and only if $\lambda^-(m,n) \leq r$. In particular, $0 < \lambda^-(m_n(r),n), \lambda^-(M_n(r),n) \leq r$. On the other hand, $\lambda^-(m,n)$ is a solution to the equation

(4.47) $$-D(\lambda^-(m,n), m, n) = -qn^{-4}.$$

By (4.40) we have

$$-\partial_m D(\lambda, m, n) = \frac{1}{n^2}, \quad \frac{\omega_1}{2} \leq -\partial_\lambda D(\lambda, m, n) \leq \frac{3\omega_1}{2} \quad \text{for } \lambda \in [0, r].$$

Hence for fixed $n \geq N_3$, the implicit function $\lambda^-(m,n)$ satisfying the equation (4.47) is defined and strongly decreases on the interval $[m_n(r), M_n(r)]$ which yields (4.36). The relation $I(k,n) \cap [0,r] \neq \emptyset$ for each integer $k \in [m_n(r), M_n(r)]$ follows from (4.36). Next if $m_n(r) \leq k - 1 \leq k \leq M_n(r)$, then we have

$$0 = -D(\lambda^-(k-1,n), k-1, n) + D(\lambda^-(k,n), k, n) =$$
$$= -D(\lambda^-(k-1,n), k-1, n) + D(\lambda^-(k,n), k-1, n) +$$
$$- D(\lambda^-(k,n), k-1, n) + D(\lambda^-(k,n), k, n)$$
$$= -D(\lambda^-(k-1,n), k-1, n) + D(\lambda^-(k,n), k-1, n) - \frac{1}{n^2}$$
$$\leq \max_{\lambda \in [0,r]} \{-\partial_\lambda D(\lambda, k-1, n)\}(\lambda^-(k-1,n) - \lambda^-(k,n)) - \frac{1}{n^2}$$
$$\leq \frac{3\omega_1}{2}(\lambda^-(k-1,n) - \lambda^-(k,n)) - \frac{1}{n^2},$$

and hence
$$\frac{3\omega_1}{2}(\lambda^-(k-1,n) - \lambda^-(k,n)) \geq \frac{1}{n^2},$$

which completes the proof of (ii).

Next note that (4.38) is a consequence of the inequalities

$$\frac{2c_0}{5\omega_1}\frac{1}{n^{3+\alpha}} \leq \frac{2}{5\omega_1}d(m,n) \leq \lambda^-(m,n) \leq r,$$

and (iii) follows.

It remains to prove (iv). We have

$$D_\infty(\lambda^\pm_\infty, m, n) - D_\infty(\lambda^\pm_j, m, n) = D_j(\lambda^\pm_j, m, n) - D_\infty(\lambda^\pm_j, m, n).$$

On the other hand,

$$|D_j(\lambda^\pm_j, m, n) - D_\infty(\lambda^\pm_j, m, n)| \leq \frac{\lambda^\pm_j}{N^2}|\varphi_\infty - \varphi_j| + (\lambda^\pm_j)^{3/2}|\Omega_\infty - \Omega_j| \leq$$
$$2r_2 R 2^{-j}(N^{-2} + 1) \leq 4r_2 R 2^{-j},$$

which along with (4.40) and (4.31) implies

$$|\lambda^\pm_\infty(m,n) - \lambda^\pm_j(m,n)| \leq \frac{8R}{\omega_1}r_2 2^{-j} \leq \frac{\omega_1}{16}2^{-j},$$

and the lemma follows. □

Cardinality of the set $\{I_j(m,n)\}$. Set

$$\theta_n = \{\omega_0 n^2 - C\} = \text{ decimal part of } \omega_0 n^2 - C,$$

and introduce the sequence of numbers $\delta_n(r)$ defined by

$$\delta_n(r) = 1 \text{ when } \theta_n \leq \frac{5}{2}\omega_1 r n^2, \text{ and } \delta_n(r) = 0 \text{ otherwise.}$$

LEMMA 4.12. *For each* $n \geq N_3$ *and* $0 < r < r_2$, $1 \leq j \leq \infty$,
(4.48) $\qquad \text{card}\{m : I_j(m,n) \cap [0,r] \neq \emptyset\} \leq \delta_n + crn^2.$

PROOF. For simplicity we omit the index j. Denote by $\mathcal{I}$ the totality of all intervals $I(m,n)$ given by Lemma 4.11 such that $I(m,n) \cap [0, r_2] \neq \emptyset$. By assertion (ii) of Lemma 4.11 there is one-to-one correspondence between the intervals $I(m,n) \in \mathcal{I}$, having nonempty intersection with $[0, r]$, and the sequence of $\lambda^-(k, n)$ given by (4.36). In particular, we have

(4.49) $$\operatorname{card}\{m : I(m,n) \cap [0,r] \neq \emptyset\} = M_n(r) - m_n(r) + 1.$$

On the other hand, inequality (4.37) yields

$$\lambda^-(m_n(r), n) - \lambda^-(M_n(r), n) \geq (M_n(r) - m_n(r))\frac{2}{3\omega_1 n^2},$$

Since $m_n(r), M_n(r) \in [0, r]$, we conclude from this that

$$M_n(r) - m_n(r) \leq cn^2 r.$$

Combining this inequality with (4.49) we obtain

(4.50) $$\operatorname{card}\{m : I(m,n) \cap [0,r] \neq \emptyset\} \leq 1 + crn^2$$

On the other hand, for given $n > N_3$,

$$\left(\bigcup_{m:I(m,n)\in\mathcal{I}} I(m,n)\right) \cap [0,r] = \emptyset \text{ if } \min_{m:I(m,n)\in\mathcal{I}} \lambda_-(m,n) > r.$$

Since, by Lemma 4.11, $0 < \frac{2}{5\omega_1}d(m,n) \leq \lambda_-(m,n)$, we conclude from this that

$$\left(\bigcup_{m:I(m,n)\in\mathcal{I}} I(m,n)\right) \cap [0,r] = \emptyset \text{ if } \min_{m:d(m,n)>0} d(m,n) > \frac{5}{2}\omega_1 r.$$

Obviously

$$\min_{m:d(m,n)>0} d(m,n) = \frac{1}{n^2}\min_{m:\omega_0 n^2 - C - m > 0}(\omega_0 n^2 - C - m) = \frac{\theta_n}{n^2},$$

which yields

(4.51) $$\operatorname{card}\{m : I(m,n) \cap [0,r] \neq \emptyset\} = 0 \text{ for } \frac{\theta_n}{n^2} > \frac{5}{2}\omega_1 r$$

Combining (4.50) and (4.51) we obtain

$$\sum_{m:I(m,n)\cap[0,r]\neq\emptyset} 1 = \operatorname{card}\{m : I(m,n) \cap [0,r] \neq \emptyset\} \leq \delta_n(r) + crn^2,$$

and the lemma follows. $\square$

Proof of Theorem 4.10. First note that

$$\operatorname{meas}\left([0,r]\backslash\bigcap_{j=1}^{\infty}\mathcal{H}_j(N,q,r)\right) \leq \sum_{\substack{(m,n) : n \geq N_3 \\ I_j(m,n) \cap [0,r] \neq \emptyset}} \operatorname{meas}\left(\bigcup_{j=1}^{\infty} I_j(m,n)\right) := \wp(r).$$

The following lemma gives an estimate for $\wp$ in terms of the sequence $\delta_n(r)$.

4.1. PROOF OF THEOREM 4.10

LEMMA 4.13. *For each $N \geq N_3$, $0 < r < r_2$, and $\sigma \in (0, 1/(3+\alpha))$,*

$$\wp(r) \leq \sum_{n \geq c_3 r^{-1/(3+\alpha)}} c \frac{\delta_n(r)}{n^4} (\ln n + 1) + c r^{1+\sigma}, \tag{4.52}$$

where c depends on ω_1, α, σ and R only, $c_3 = \left(\frac{2c_0}{5\omega_1}\right)^{1/(3+\alpha)}$, c_0 is the constant from Condition M.

PROOF. Introduce the sequence

$$\varsigma(n) = \frac{4}{\ln 2} \ln n - \frac{1}{\ln 2} \ln \left[\frac{64}{3\omega_1^2} q\right], \quad n \geq 1.$$

It is easy to see that $\omega_1 2^{-j-4} < 4q/(3\omega_1 n^4)$ for all $j > \varsigma(n)$. It follows from this and (4.39) that for all such j, the intervals $I_j(m, n)$ have nonempty intersections with $I_\infty(m, n)$ and $\cup_{j > \varsigma(n)} I_j(m, n) \subset \tilde{I}_\infty(m, n)$, which yields

$$\text{meas}\left(\bigcup_{j > \varsigma(n)} I_j(m, n)\right) \leq \text{meas } \tilde{I}_\infty(m, n) \leq \frac{20q}{3\omega_1} \frac{1}{n^4},$$

where $\tilde{I}_\infty(m, n)$ is the $4q/(3\omega_1 n^4)$-neighborhood of $I_\infty(m, n)$. Thus we get

$$\wp(r) \leq \sum_{\substack{n \geq N_3 \\ I_j(m,n) \cap [0,r] \neq \emptyset}} \sum_{j=1}^{\varsigma(n)} \sum_{(m: I_j(m,n) \cap [0,r] \neq \emptyset)} \text{meas } I_j(m, n) +$$

$$\sum_{\substack{n \geq N_3 \\ I_j(m,n) \cap [0,r] \neq \emptyset}} \sum_{(m: I_\infty(m,n) \cap [0,r] \neq \emptyset)} \text{meas}\left(\bigcup_{j > \varsigma(n)} I_j(m, n)\right),$$

which leads to

$$\wp(r) \leq \sum_{\substack{n \geq N_3 \\ I_j(m,n) \cap [0,r] \neq \emptyset}} \frac{c}{n^4} \sum_{j=1}^{\varsigma(n)} \sum_{(m: I_j(m,n) \cap [0,r] \neq \emptyset)} 1 +$$

$$\sum_{n \geq N_3} \frac{c}{n^4} \sum_{(m: I_\infty(m,n) \cap [0,r] \neq \emptyset)} 1,$$

Since, by Lemma 4.11, the intersection $I_j(m, n) \cap [0, r]$ is empty for $n < c_3 r^{-1/(3+\alpha)}$, we have

$$\wp(r) \leq \sum_{n \geq c_3 r^{-1/(3+\alpha)}} \frac{c}{n^4} \sum_{j=1}^{\varsigma(n)} \text{card } \{m : I_j(m, n) \cap [0, r] \neq \emptyset\} +$$

$$\sum_{n \geq c_3 r^{-1/(3+\alpha)}} \frac{c}{n^4} \text{card } \{m : I_\infty \cap [0, r] \neq \emptyset\},$$

which along with (4.48) yields

$$\wp(r) \leq \sum_{n \geq c_3 r^{-1/(3+\alpha)}} \frac{c}{n^4} \varsigma(n)(\delta_n(r) + crn^2) + \sum_{n \geq c_3 r^{-1/(3+\alpha)}} \frac{c}{n^4}(\delta_n + crn^2),$$

$$\leq c \sum_{n \geq c_3 r^{-1/(3+\alpha)}} \frac{\delta_n(r)}{n^4}(1 + \ln n) + cr \sum_{n \geq c_3 r^{-1/(3+\alpha)}} \frac{1 + \ln n}{n^2}$$

$$\leq \sum_{n \geq c_3 r^{-1/(3+\alpha)}} c \frac{\delta_n(r)}{n^4}(1 + \ln n) + cr^{1+\sigma},$$

which completes the proof. □

Introduce the sequence

$$\nu_n(r) = 1 \text{ when } \theta_n \leq \frac{5}{2}\omega_1 r^{\frac{1-\alpha}{3-\alpha}}, \text{ and } \nu_n(r) = 0 \text{ otherwise.}$$

LEMMA 4.14. *Let* $0 < r \leq r_2$ *and* $0 < \beta < \alpha/(3-\alpha)$. *If* $c_3 r^{\frac{-1}{3+\alpha}} \geq r^{\frac{-1}{3-\alpha}}$, *then*

(4.53) $$\wp(r) \leq cr^{1+\beta}.$$

If $c_3 r^{\frac{-1}{3+\alpha}} < r^{\frac{-1}{3-\alpha}}$, *then*

(4.54) $$\wp(r) \leq c \sum_{c_3 r^{-1/(3+\alpha)} \leq n \leq r^{-1/(3-\alpha)}} \frac{\nu_n(r)}{n^4}(1 + \ln n) + cr^{1+\beta},$$

where c does not depend on r.

PROOF. Since $\beta < 1/(3+\alpha)$, we have from inequality (4.52)

(4.55) $$\wp(r) \leq c\Pi(r) + \sum_{n \geq r^{-1/(3-\alpha)}} c \frac{\delta_n(r)}{n^4}(1 + \ln n) + cr^{1+\beta},$$

where

$$\Pi(r) = \sum_{c_3 r^{-1/(3+\alpha)} \leq n \leq r^{-1/(3-\alpha)}} \frac{\delta_n(r)}{n^4}(1 + \ln n) \text{ for } c_3 r^{\frac{-1}{3+\alpha}} < r^{\frac{-1}{3-\alpha}},$$

and $\Pi(r) = 0$ otherwise. It is easy to see that

(4.56) $$\sum_{n \geq r^{-1/(3-\alpha)}} \frac{\delta_n(r)}{n^4}(1 + \ln n) \leq \sum_{n \geq r^{-1/(3-\alpha)}} \frac{1}{n^4}(1 + \ln n) \leq cr^{1+\beta}$$

If $n \leq r^{-1/(3-\alpha)}$ then $\theta_n > \frac{5}{2}\omega_1 r^{(1-\alpha)/(3-\alpha)}$ yields $\theta_n > \frac{5}{2}\omega_1 n^2 r$. In other words, equality $\nu_n(r) = 0$ yields $\delta_n(r) = 0$. Hence

(4.57) $$\Pi(r) \leq \sum_{c_3 r^{-1/(3+\alpha)} \leq n \leq r^{-1/(3-\alpha)}} \frac{\nu_n(r)}{n^4}(1 + \ln n) \text{ for } c_3 r^{\frac{-1}{3+\alpha}} < r^{\frac{-1}{3-\alpha}},$$

and $\Pi(r) = 0$ otherwise. Substituting (4.56) and (4.57) into (4.55) and noting that we obtain needed inequalities (4.53), (4.54). □

Now set
$$r_3 = \min\{((\tfrac{5}{2}\omega_1)^{-1}c_3^{-\tfrac{1}{78}})^{1/\iota}, c_3^{3+\alpha}\left(\frac{1}{4}\right)^{78(3+\alpha)}\}, \quad \iota = \frac{1-\alpha}{3-\alpha} - \frac{1}{78(3+\alpha)}.$$

Remark This complicated formulae simply express the fact that the number $\rho = \left(c_3^{-1}r^{1/(3+\alpha)}\right)^{1/78}$ satisfies the inequalities
$$\tfrac{5}{2}\omega_1 r^{(1-\alpha)/(3-\alpha)} \leq \rho \leq 1/4 \text{ for } r \leq r_3.$$

LEMMA 4.15. *Let , $0 < r \leq r_3$, $c_3 r^{-1/(3+\alpha)} < r^{-1/(3-\alpha)}$, and $0 < \gamma < (1/78 - \alpha)/(3+\alpha)$. Then*

(4.58) $$\sum_{c_3 r^{-1/(3+\alpha)} \leq n \leq r^{-1/(3-\alpha)}} \frac{\nu_n(r)}{n^4}(1 + \ln n) \leq cr^{1+\gamma}.$$

PROOF. Let p and q be the minimal and maximal integers from the interval $(c_3 r^{-1/(3+\alpha)}, r^{-1/(3-\alpha)})$. Introduce the average
$$S_n(r) = \frac{1}{n}\sum_{k=1}^{n} \nu_k(r),$$
Noting that $\nu_n(r) = nS_n - (n-1)S_{n-1}$, we obtain
$$\sum_{c_3 r^{-1/(3+\alpha)} \leq n \leq r^{-1/(3-\alpha)}} \frac{\nu_n(r)}{n^4} = \sum_{p \leq n \leq q} \frac{\nu_n(r)}{n^4}$$
$$\leq \sum_{p \leq n \leq q-1}\left(\frac{1+\ln n}{n^4} - \frac{1+\ln(n+1)}{(n+1)^4}\right)nS_n + \frac{1}{q^3}S_q.$$

Since $q \geq c_3 r^{-1/(3+\alpha)}$ and
$$\left(\frac{1+\ln n}{n^4} - \frac{1+\ln(n+1)}{(n+1)^4}\right)n \leq \frac{c}{n^4}(1+\ln n),$$
we conclude from this that
$$\sum_{c_3 r^{-1/(3+\alpha)} \leq n \leq r^{-1/(3-\alpha)}} \frac{\nu_n(r)}{n^4}(1+\ln n)$$
$$\leq c\{\sum_{c_3 r^{-1/(3+\alpha)} \leq n \leq r^{-1/(3-\alpha)}} \frac{1}{n^4}(1+\ln n) + r^{3/(3+\alpha)}\} \sup_{n \geq c_3 r^{-1/(3+\alpha)}} S_n \leq$$

(4.59) $$cr^{(3-\vartheta)/(3+\alpha)} \sup_{n \geq c_3 r^{-1/(3+\alpha)}} S_n,$$

where ϑ is an arbitrary positive number. Now set
$$\rho = \left(c_3^{-1}r^{1/(3+\alpha)}\right)^{1/78}.$$

It follows from the choice of r_3 and the inequality $r < r_3$ that

(4.60) $$\tfrac{5}{2}\omega_1 r^{(1-\alpha)/(3-\alpha)} \leq \rho < 1/4, \quad n \geq c_3 r^{-1/(3+\alpha)} \Rightarrow n \geq \rho^{-78}.$$

It follows from this that
$$S_n \equiv \frac{1}{n} \sum_{\substack{1 \leq k \leq n \\ \theta_n \leq \frac{5}{2}\omega_1 r^{(1-\alpha)/(3-\alpha)}}} 1 \leq \frac{1}{n} \sum_{\substack{1 \leq k \leq n, \\ \theta_n \in [0, \rho]}} 1.$$

Applying Proposition E.1 we obtain that for any $n \geq c_3 r^{-1/(3+\alpha)} = \rho^{-78}$
$$S_n \leq c\rho = cr^{1/78(3+\alpha)},$$

Combining this inequality with (4.59) we finally obtain
$$\sum_{c_3 r^{-1/(3+\alpha)} \leq n \leq r^{-1/(3-\alpha)}} \frac{\nu_n(r)}{n^4}(1 + \ln n) \leq cr^{1/78(3+\alpha)} r^{(3-\vartheta)/(3+\alpha)} = cr^{1+\gamma},$$

and the lemma follows. □

Finally, combining inequalities (4.53), (4.54) and (4.58) we conclude that for $0 < r < r_3$,
$$\wp(r) \leq cr(r^\beta + r^\gamma) \leq cr^{1+\varpi},$$
which completes the proof of Theorem 4.10.

CHAPTER 5

Descent Method-Inversion of the Linearized Operator

In this chapter we give a general method of reduction, the descent method, allowing to transform the original linear operator of order 2 into the sum of a main operator with constant coefficients and a smoothing perturbation operator.

Let us consider the basic operator equation

(5.1) $$\mathfrak{L}u + \mathfrak{A}\mathfrak{D}_1 u + \mathfrak{B}u + \mathfrak{L}_{-1}u = f,$$

with zero-order pseudodifferential operators $\mathfrak{A}$, $\mathfrak{B}$ and an integro-differential operator $\mathfrak{L}_{-1}$ of order -1. Assume that they satisfy the following conditions.

Symmetry condition:

(i) For all $Y \in \mathbb{T}^2$ and $|\xi| \leq 1$,

(5.2) $$A(Y, -\xi) = \overline{A(Y, \xi)},$$

which means in particular that $\mathfrak{A}u$ is real for real-valued functions u.

(ii) Equation (5.1) is invariant with respect to the symmetry $Y \to Y^* = (-y_1, y_2)$. This is equivalent to the equivariant property

(5.3) $$\begin{aligned} \mathfrak{A}\mathfrak{D}_1 u(Y^*) &= \mathfrak{A}\mathfrak{D}_1 u^*(Y), \quad \mathfrak{B}u(Y^*) = \mathfrak{B}u^*(Y), \\ \mathfrak{L}_{-1}u(Y^*) &= \mathfrak{L}_{-1}u^*(Y), \quad u^*(Y) = u(Y^*), \end{aligned}$$

which can be also written in the form

(5.4) $$A(Y^*, \xi^*) = -A(Y, \xi), \quad B(Y^*, \xi^*) = B(Y, \xi).$$

(iii) For each $s \in [1, l-3]$, $H_{o,e}^s(\mathbb{R}^2/\Gamma)$ is invariant subspace of operators $\mathfrak{A}\mathfrak{D}_1$, $\mathfrak{B}$ and $\mathfrak{L}_{-1}$.

Metric condition:

(iv) There are exponents r, s, and l, satisfying inequalities

$$1 \leq r \leq s \leq l - 10,$$

and $\varepsilon \in (0, 1]$ so that

(5.5) $$|\mathfrak{A}|_{4,r} + |\mathfrak{B}|_{4,r} \leq \varepsilon, \quad |\mathfrak{A}|_{4,l} + |\mathfrak{B}|_{4,l} < \infty.$$

For each $s \in [r, l-10]$ there is a constant C_s so that

(5.6) $$\|\mathfrak{L}_{-1}u\|_r \leq c\varepsilon\|u\|_{r-1},$$
$$\|\mathfrak{L}_{-1}u\|_s \leq c\big(\varepsilon\|u\|_{s-1} + C_s\|u\|_0\big).$$

Restrictions on the spectrum and resolvent of $\mathfrak{L}$:

(v) The selfadjoint operator $\mathfrak{L} : H^s_{o,e}(\mathbb{R}^2/\Gamma) \mapsto H^s_{o,e}(\mathbb{R}^2/\Gamma)$ has a simple eigenvalue $\nu_1 \varepsilon^2 + O(\varepsilon^3)$ with corresponding eigenfunction $\psi^{(0)}$; the space $H^s_{o,e}(\mathbb{R}^2/\Gamma)$ is the sum of orthogonal subspaces

$$H^s_{o,e}(\mathbb{R}^2/\Gamma) = \text{span } \{\psi^{(0)}\} \oplus H^{s,\perp}_{o,e};$$

$H^{s,\perp}_{o,e}$ are invariant subspaces of the operator $\mathfrak{L}$. Denote by $\mathcal{Q}$ the orthogonal projector of $H^s_{o,e}(\mathbb{R}^2/\Gamma)$ onto $H^{s,\perp}_{o,e}$.

(vi) For

$$(5.7) \qquad \varkappa = \frac{1}{16\nu\pi^2} \int_{\mathbb{T}^2} \Big(A(Y,0,1)^2 - 4\nu B(Y,0,1)\Big) dY,$$

and all $s \geq 1$, the inverse $(\mathfrak{L} - \varkappa)^{-1} : H^{s,\perp}_{o,e} \mapsto H^{s-1,\perp}_{o,e}$ is continuous and

$$(5.8) \qquad \|(\mathfrak{L} - \varkappa)^{-1} u\|_{s-1} \leq c(s) \|u\|_s.$$

Nondegeneracy condition:

(vii)
(5.9)
$$\varkappa = O(\varepsilon^2), \quad \int_{\mathbb{T}^2} \Big((\mathfrak{L} + \mathfrak{H})\psi^{(0)} - \mathfrak{H}\mathcal{Q}(\mathfrak{L}-\varkappa)^{-1}\mathcal{Q}\mathfrak{H}\psi^{(0)}\Big)\psi^{(0)} \, dY = @\varepsilon^2 + O(\varepsilon^3),$$

where @ is a non-zero absolute constant, and the operator $\mathfrak{H} = \mathfrak{A}\mathfrak{D}_1 + \mathfrak{B} + \mathfrak{L}_{-1}$.

Here and below we denote by c generic constants depending on r and l only, and use the standard notation $O(\varepsilon^n)$ for quantities which absolute value does not exceed $c\varepsilon^n$. The following theorem is the main result of this chapter.

THEOREM 5.1. *Under the above assumptions, there is a positive constant ε_0 depending on l, r only so that for any $f \in H^s_{o,e}(\mathbb{R}^2/\Gamma)$ and $\varepsilon < \varepsilon_0$, equation (5.1) has a unique solution $u \in H^{s-1}_{o,e}(\mathbb{R}^2/\Gamma)$ satisfying the inequalities*

$$(5.10) \qquad \|u\|_{r-1} \leq \varepsilon^{-2} c \|f\|_r,$$
$$(5.11) \qquad \|u\|_{s-1} \leq \varepsilon^{-2} c \|f\|_r \big(1 + C_s + |\mathfrak{A}|_{4,s+10} + |\mathfrak{B}|_{4,s+10}\big) + c\|f\|_s.$$

PROOF. The proof constitutes the next two sections. □

5.1. Descent method

In this section we develop an algebraic method which allows us to reduce (5.1) to a Fredholm-type equation. The main result in this direction is the following

THEOREM 5.2. *Let zero-order pseudodifferential operators $\mathfrak{A}$ and $\mathfrak{B}$ satisfy symmetry conditions (i), (ii), metric condition (iv), and $\varkappa$ is given* by (5.7). Then there exist integro-differential operators $\mathfrak{C}$, $\mathfrak{E}$, $\mathfrak{F}$ satisfying symmetry condition (5.3) with $\mathfrak{B}$ replaced by $\mathfrak{C}$, $\mathfrak{E}$, $\mathfrak{F}$ so that

$$(5.12) \qquad \left(\mathfrak{L} + \mathfrak{A}\mathfrak{D}_1 + \mathfrak{B}\right)(1 + \mathfrak{C})\Pi_1 = (1 + \mathfrak{E})\Pi_1(\mathfrak{L} - \varkappa) + \mathfrak{F},$$

and for $1 \leq s \leq l - 10$,

$$(5.13) \quad \begin{aligned} \|\mathfrak{C}u\|_s + \|\mathfrak{E}u\|_s &\leq c\varepsilon\|u\|_s + c\big(|\mathfrak{A}|_{4,s+9} + |\mathfrak{B}|_{4,s+9}\big)\|u\|_0, \\ \|\mathfrak{F}u\|_s &\leq c\varepsilon\|u\|_{s-1} + c\big(|\mathfrak{A}|_{4,s+10} + |\mathfrak{B}|_{4,s+10}\big)\|u\|_0. \end{aligned}$$

The proof falls into three steps and is based on the following proposition, which gives the special decomposition of zero-order operators and plays the key role in our analysis. In order to formulate it we introduce the important notion of an elementary operator.

DEFINITION 5.3. Let $\tilde{W} : \mathbb{T}^2 \mapsto \mathbb{C}$ be a function of class $C^l(\mathbb{R}^2/\Gamma)$. We say that $\mathfrak{W}$ is the elementary operator associated with $\tilde{W}$, if $\mathfrak{W}$ is a zero-order pseudodifferential operator with the symbol $W(Y, \xi_2) = \operatorname{Re} \tilde{W}(Y) + i\xi_2 \operatorname{Im} \tilde{W}(Y)$.

PROPOSITION 5.4. Let a zero-order pseudodifferential operators $\mathfrak{A}$ with symbol $A(Y,\xi)$, and an elementary operator $\mathfrak{W}$, satisfy conditions (3.12), (5.2), and $\mathfrak{S}$ be a pseudodifferential operator with the symbol $S(Y,\xi) = W(Y,\xi_2)A(Y,\xi)$. Then there exist pseudodifferential operators $\mathfrak{M}_A$, $\mathfrak{P}_A$, $\mathfrak{U}_S$ and $\mathfrak{N}_A$, $\mathfrak{Q}_A$, $\mathfrak{V}_S$ so that

$$(5.14) \qquad \mathfrak{A}\Pi_1 = \sum_{j=0}^{1} \mathfrak{A}_j \mathfrak{D}_1^{-j} + \mathfrak{M}_A \mathfrak{L} + \mathfrak{N}_A,$$

$$(5.15) \qquad \mathfrak{A}\mathfrak{D}_1 = \sum_{j=0}^{2} \mathfrak{A}_j \mathfrak{D}_1^{1-j} + \mathfrak{P}_A \mathfrak{L} + \mathfrak{Q}_A,$$

$$(5.16) \qquad \mathfrak{S}\mathfrak{D}_1 = \sum_{j=0}^{2} \mathfrak{S}_j \mathfrak{D}_1^{1-j} + \mathfrak{U}_S \mathfrak{L} + \mathfrak{V}_S.$$

Here $\mathfrak{A}_j$, $\mathfrak{S}_j$ are elementary pseudodifferential operators associated with the complex-valued functions

$$(5.17) \qquad \tilde{A}_0(Y) = A(Y, 0, 1), \quad \tilde{A}_1(Y) = \frac{1}{\nu}[\partial_{\xi_1} A](Y, 0, 1),$$

$$(5.18) \qquad \tilde{A}_2(Y) = \frac{1}{2\nu^2}[(\partial_{\xi_1}^2 - \partial_{\xi_2})A + i\operatorname{Im} A](Y, 0, 1),$$

$$(5.19) \qquad \tilde{S}_j = \tilde{A}_j \tilde{W} \text{ for } j = 0, 1, \quad \tilde{S}_2 = \tilde{A}_2 \tilde{W} - \left(\frac{1}{\nu}\right)^2 \operatorname{Im} \tilde{A}_0 \operatorname{Im} \tilde{W}.$$

For any $u \in H^s(\mathbb{R}^2/\Gamma)$ with $1 \leq s < l - 4$,

(5.20)
$$\|\mathfrak{M}_A u\|_s + \|\mathfrak{N}_A u\|_s + \|\mathfrak{P}_A u\|_s + \|\mathfrak{Q}_A u\|_s \leq$$
$$c(s)\Big(|\mathfrak{A}|_{3,s}\|u\|_0 + |\mathfrak{A}|_{3,3}\|u\|_{s-1}\Big),$$

(5.21)
$$\|\mathfrak{U}_S u\|_{s-1} + \|\mathfrak{V}_S u\|_s \leq$$
$$c(s)\Big(\big(|\mathfrak{A}|_{3,3}|\mathfrak{W}|_{3,s+3} + |\mathfrak{A}|_{3,s+3}|\mathfrak{W}|_{3,3}\big)\|u\|_0 + |\mathfrak{A}|_{3,3}|\mathfrak{W}|_{3,3}\|u\|_{s-1}\Big)$$

Proof: The proof is given in Appendix F. Let us turn to the proof of Theorem 5.2.

First step

We begin with the calculation of a commutator of an elementary operator and the left side of (5.12).

LEMMA 5.5. *Let $\mathfrak{W}$ be the elementary operator associated with a function $\tilde{W} \in C^l(\mathbb{R}^2/\Gamma)$. Then*

(5.22)
$$\Big(\mathfrak{L} + \mathfrak{A}\mathfrak{D}_1 + \mathfrak{B}\Big)\mathfrak{W} = \mathfrak{W}\mathfrak{L} + \Big(\mathfrak{W}_1 + \mathfrak{S}\Big)\mathfrak{D}_1 + \mathfrak{T} + \mathfrak{R},$$

where symbols of an elementary operator $\mathfrak{W}_1$ and zero-order pseudodifferential operators $\mathfrak{S}, \mathfrak{T}$ are given by

(5.23)
$$W_1 = 2\nu \partial_{y_1} W, \quad S = AW,$$
$$T = \nu \partial_{y_1}^2 W - i\xi_1 \partial_{y_1} W - i\tau \xi_2 \partial_{y_2} W + A \partial_{y_1} W + BW +$$
$$\xi_1(\xi^\perp \cdot \partial_\xi A)(\xi_2 \partial_{y_1} - \tau \xi_1 \partial_{y_2})W,$$

where $\xi^\perp = (\xi_2, -\xi_1)$. The remainder has the estimate

(5.24)
$$\|\mathfrak{R} u\|_s \leq c(s)\Big(|1 - \mathfrak{W}|_{0,6}\|u\|_{s-1} + |1 - \mathfrak{W}|_{0,s+6}\|u\|_0\Big) +$$
$$c(s)\Big(|\mathfrak{W}|_{0,6}|\mathfrak{A}|_{2,3}\|u\|_{s-1} + (|\mathfrak{W}|_{0,s+6}|\mathfrak{A}|_{2,3} + |\mathfrak{W}|_{0,6}|\mathfrak{A}|_{2,s})\|u\|_0\Big) +$$
$$c(s)\Big(|\mathfrak{W}|_{0,4}|\mathfrak{B}|_{2,3}\|u\|_{s-1} + (|\mathfrak{W}|_{0,s+4}|\mathfrak{B}|_{2,3} + |\mathfrak{W}|_{0,5}|\mathfrak{B}|_{2,s})\|u\|_0\Big).$$

PROOF. It is easy to see that

(5.25)
$$\nu \mathfrak{D}_1^2 \mathfrak{W} = \nu \mathfrak{W} \mathfrak{D}_1^2 + \mathfrak{W}_1 \mathfrak{D}_1 + \mathfrak{W}_2,$$

where $\mathfrak{W}_2$ is the elementary operator associated with the functions $\nu \partial_{y_1}^2 \tilde{W}$. Next since $(-\Delta)^{-1/2}$ and $\mathfrak{W}$ are first and zero order pseudodifferential operators, formulae for commutators (F.6), (F.8) from Proposition F.3 imply

(5.26)
$$(-\Delta)^{1/2}\mathfrak{W} = \mathfrak{W}(-\Delta)^{1/2} + [(-\Delta)^{1/2}, \mathfrak{W}]_1 + \mathfrak{D}^{[N,W]},$$

where the symbol of the operator $[(-\Delta)^{1/2}, \mathfrak{W}]_1$ is equal to

(5.27)
$$-i(\xi_1 \partial_{y_1} + \tau \xi_2 \partial_{y_2})W.$$

5.1. DESCENT METHOD

Moreover, since $|1 - \mathfrak{W}|_{m,l} \leq \|1 - \tilde{W}\|_{C^l}$ and the symbol of the operator $(-\Delta)^{1/2}$ does not depend on Y, inequality (F.10) yields the estimate

$$(5.28) \qquad \|\mathfrak{D}^{[N,W]}u\|_s \leq c(s)\big(\|1 - \tilde{W}\|_{C^{6+s}}\|u\|_0 + \|1 - \tilde{W}\|_{C^6}\|u\|_{s-1}\big).$$

Next applying formulae (F.5),(F.7) to the composition of the pseudodifferential operators $\mathfrak{A}\mathfrak{D}_1$ and $\mathfrak{W}$ we arrive at the equality

$$(5.29) \qquad \mathfrak{A}\mathfrak{D}_1\mathfrak{W} = \mathfrak{S}\mathfrak{D}_1 + (\mathfrak{A}\mathfrak{D}_1\mathfrak{W})_1 + \mathfrak{D}^{(AD_1W)},$$

where $(\mathfrak{A}\mathfrak{D}_1\mathfrak{W})_1$ is a zero order pseudodifferential operator which symbol is given by formula (F.7) with A replaced by $ik_1 A$ and B replaced by W. Noting that for $k \neq 0$,

$$k_1 \partial_{k_1} \xi = \xi_1 \xi_2 \xi^\perp, \quad k_1 \partial_{k_2} \xi = -\tau \xi_1^2 \xi^\perp \text{ with } \xi^\perp = (\xi_2, -\xi_1)$$

we can rewrite expression (F.7) for the symbol $(\mathfrak{A}\mathfrak{D}_1\mathfrak{W})_1$ in the form

$$(5.30) \qquad A\partial_{y^1} W + \xi_1 \big[\xi^\perp \cdot \partial_\xi A\big] \big[\xi_2 \partial_{y_1} W - \tau \xi_1 \partial_{y_2} W\big].$$

Recall that it vanishes for $k = 0$. Since $\mathfrak{A}\mathfrak{D}_1$ is a first order operator with $|\mathfrak{A}\mathfrak{D}_1|_{1,l}^1 \leq c|\mathfrak{A}|_{1,l}$ inequality (F.9) from Proposition F.3 implies the estimate

$$(5.31) \qquad \|\mathfrak{D}^{AD_1W}u\|_s \leq c\big(|\mathfrak{A}|_{2,s}|\mathfrak{W}|_{0,6} + |\mathfrak{A}|_{2,3}|\mathfrak{W}|_{0,6+s}\big)\|u\|_0 + c|\mathfrak{A}|_{2,3}|W|_{0,6}\|u\|_{s-1}.$$

Setting

$$\mathfrak{T} = \mathfrak{W}_2 + [(-\Delta)^{1/2}, \mathfrak{W}]_1 + (\mathfrak{A}\mathfrak{D}_1\mathfrak{W})_1 + (\mathfrak{B}\mathfrak{W})_0$$

we obtain the needed representation. with the remainder

$$\mathfrak{D}^{[N,W]} + \mathfrak{D}^{(AD_1W)} + \mathfrak{D}^{(BW)},$$

and the lemma follows. $\square$

Second step

Now we give a formal construction of the operators $\mathfrak{C}$, $\mathfrak{E}$ and $\mathfrak{F}$. We take the operator $\mathfrak{C}$ in the form

$$(5.32) \qquad \mathfrak{C} = \sum_{p=0}^{2} \mathfrak{W}^{(p)} \mathfrak{D}_1^{-p} - \Pi_1,$$

where elementary operators $\mathfrak{W}^{(p)}$ will be specified below. Applying Lemma 5.5 and noting that $\mathfrak{L}$ commutes with $\mathfrak{D}_1^j$ we obtain

$$(\mathfrak{L} + \mathfrak{A}\mathfrak{D}_1 + \mathfrak{B})\mathfrak{W}^{(p)}\mathfrak{D}_1^{-p} = \mathfrak{W}^{(p)}\mathfrak{D}_1^{-p}\mathfrak{L} + \mathfrak{W}_1^{(p)}\mathfrak{D}_1^{1-p} + \\ + \mathfrak{S}^{(p)}\mathfrak{D}_1^{1-p} + \mathfrak{T}^{(p)}\mathfrak{D}_1^{-p} + \mathfrak{R}^{(p)}\mathfrak{D}_1^{-p}.$$

Here $\mathfrak{S}^{(p)}$, $\mathfrak{T}^{(p)}$, $\mathfrak{R}^{(p)}$ are given by Lemma 5.5 with $\mathfrak{W}$ replaced by $\mathfrak{W}^{(p)}$. Combining these identities we arrive at

(5.33)
$$\big(\mathfrak{L} + \mathfrak{A}\mathfrak{D}_1 + \mathfrak{B}\big)(\Pi_1 + \mathfrak{C}) = (\Pi_1 + \mathfrak{C})\mathfrak{L} + \\ + \sum_{p=0}^{2}\Big(\mathfrak{W}_1^{(p)}\mathfrak{D}_1^{1-p} + \mathfrak{S}^{(p)}\mathfrak{D}_1^{1-p} + \mathfrak{T}^{(p)}\mathfrak{D}_1^{-p} + \mathfrak{R}^{(p)}\mathfrak{D}_1^{-p}\Big).$$

Recall that $\mathfrak{S}^{(p)}$ and $\mathfrak{T}^{(p)}$ are zero-order pseudodifferential operators, hence by Proposition 5.4, they have the decomposition

$$\mathfrak{S}^{(p)}\mathfrak{D}_1 = \sum_{j=0}^{2}\mathfrak{S}_j^{(p)}\mathfrak{D}_1^{1-j} + \mathfrak{U}_{S^{(p)}}\mathfrak{L} + \mathfrak{V}_{S^{(p)}},$$

$$\mathfrak{T}^{(p)}\Pi_1 = \sum_{j=0}^{1}\mathfrak{T}_j^{(p)}\mathfrak{D}_1^{1-j} + \mathfrak{M}_{T^{(p)}}\mathfrak{L} + \mathfrak{N}_{T^{(p)}},$$

in which the symbols of elementary operators $\mathfrak{S}_j^{(p)}$ are given by the formulae (5.19) with W replaced by $W^{(p)}$, and the symbols of elementary operators $\mathfrak{T}_j^{(p)}$ are given by formulae (5.17), (5.18) with A replaced by T. Substituting these relations into (5.33) we obtain the identity

(5.34)
$$\big(\mathfrak{L} + \mathfrak{A}\mathfrak{D}_1 + \mathfrak{B}\big)(\Pi_1 + \mathfrak{C}) = (\Pi_1 + \mathfrak{C})(\mathfrak{L} - \varkappa) + \\ \sum_{p=0}^{3}\mathfrak{J}_p\mathfrak{D}_1^{1-p} + \mathfrak{R}'\mathfrak{L} + \mathfrak{R}'',$$

where the elementary operators $\mathfrak{J}_j$ and the reminders $\mathfrak{R}'$, $\mathfrak{R}''$ are given by

(5.35)
$$\mathfrak{J}_0 = \mathfrak{W}_1^{(0)} + \mathfrak{S}_0^{(0)},$$
$$\mathfrak{J}_1 = \mathfrak{W}_1^{(1)} + \mathfrak{S}_0^{(1)} + \mathfrak{S}_1^{(0)} + \mathfrak{T}_0^{(0)} + \varkappa\mathfrak{W}^{(0)},$$
$$\mathfrak{J}_2 = \mathfrak{W}_1^{(2)} + \mathfrak{S}_0^{(2)} + \mathfrak{S}_2^{(0)} + \mathfrak{S}_1^{(1)} + \mathfrak{T}_1^{(0)} + \mathfrak{T}_0^{(1)} + \varkappa\mathfrak{W}^{(1)},$$
$$\mathfrak{J}_3 = \mathfrak{S}_2^{(1)} + \mathfrak{S}_1^{(2)} + \mathfrak{S}_2^{(2)}\mathfrak{D}_1^{-1} + \mathfrak{T}_1^{(1)} + \mathfrak{T}_0^{(2)} + \mathfrak{T}_1^{(2)}\mathfrak{D}_1^{-1} + \varkappa\mathfrak{W}^{(2)},$$

(5.36)
$$\mathfrak{R}' = \sum_{p=0}^{2}\mathfrak{U}_{S^{(p)}}\mathfrak{D}_1^{-p} + \sum_{p=0}^{1}\mathfrak{M}_{T^{(p)}}\mathfrak{D}_1^{-p},$$
$$\mathfrak{R}'' = \sum_{p=0}^{2}\mathfrak{V}_{S^{(p)}}\mathfrak{D}_1^{-p} + \sum_{p=0}^{1}\mathfrak{N}_{T^{(p)}}\mathfrak{D}_1^{-p} + \sum_{p=0}^{2}\mathfrak{R}^{(p)}\mathfrak{D}_1^{-p}.$$

Noting that
$$\mathfrak{D}_1^{-2} = (-\Delta)^{-1/2}\mathfrak{D}_1^{-2}\mathfrak{L} - \nu(-\Delta)^{-1/2}.$$

we can rewrite (5.34) in the form

(5.37) $$\left(\mathfrak{L} + \mathfrak{A}\mathfrak{D}_1 + \mathfrak{B}\right)(\Pi_1 + \mathfrak{C}) = (\Pi_1 + \mathfrak{E})(\mathfrak{L} - \varkappa) + \sum_{p=0}^{2} \mathfrak{I}_p \mathfrak{D}_1^{1-p} + \mathfrak{F}$$

with

(5.38) $$\begin{aligned} \mathfrak{E} &= \mathfrak{C} + \mathfrak{R}' + \mathbf{I}_3(-\Delta)^{-1/2}\mathfrak{D}_1^{-2}, \\ \mathfrak{F} &= \mathfrak{R}'' - \nu\mathfrak{I}_3(-\Delta)^{-1/2} + \varkappa(\mathfrak{R}' + \mathfrak{I}_3(-\Delta)^{-1/2}\mathfrak{D}_1^{-2}. \end{aligned}$$

Now our task is to show that $\mathfrak{I}_j$, $j \leq 2$, vanish for an appropriate choice of elementary operators $\mathfrak{W}^{(p)}$. Note that the operator equations $\mathfrak{I}_j = 0$ are equivalent to the scalar equations $\widetilde{I}_j(Y) = 0$, in which the complex-valued functions $\widetilde{I}_j$ are associated with the operators $\mathfrak{I}_j$. This observation along with formulae (5.35) leads to the equations

(5.39) $$\begin{aligned} \tilde{W}_1^{(0)} + \tilde{S}_0^{(0)} &= 0, \\ \tilde{W}_1^{(1)} + \tilde{S}_0^{(1)} + \tilde{S}_1^{(0)} + \tilde{T}_0^{(0)} + \kappa\tilde{W}^{(0)} &= 0, \\ \tilde{W}_1^{(2)} + \tilde{S}_0^{(2)} + \tilde{S}_2^{(0)} + \tilde{S}_1^{(1)} + \tilde{T}_1^{(0)} + \tilde{T}_0^{(1)} + \varkappa\tilde{W}^{(1)} &= 0 \end{aligned}$$

By Proposition 5.4, we have $\tilde{T}_0^{(p)} = T^{(p)}(Y, 0, 1)$ and $\tilde{T}_1^{(p)} = \nu^{-1}[\partial_{\xi_1} T^{(p)}](Y, 0, 1)$, which along with equality (5.23) yields

$$\begin{aligned} \tilde{T}_0^{(p)} &= \nu\partial_{y_1}^2 \tilde{W}^{(p)} - i\tau\partial_{y_2}\tilde{W}^{(p)} + \tilde{A}_0\partial_{y_1}\tilde{W}^{(p)} + \tilde{B}_0\tilde{W}^{(p)}, \\ \tilde{T}_1^{(p)} &= -i\frac{1}{\nu}\partial_{y_1}\tilde{W}^{(p)} + 2\tilde{A}_1\partial_{y_1}\tilde{W}^{(p)} + \tilde{B}_1\partial_{y_1}\tilde{W}^{(p)}. \end{aligned}$$

Substituting these identities into (5.39) and using $\tilde{S}_j^{(p)} = \tilde{A}_j\tilde{W}^{(p)}$ we obtain the recurrent system of ordinary differential equations for functions $\tilde{W}^{(p)}$, $p = 0, 1, 2$,

(5.40) $$(2\nu\partial_{y_1} + \tilde{A}_0)\tilde{W}^{(0)} = 0,$$

(5.41) $$(2\nu\partial_{y_1} + \tilde{A}_0)\tilde{W}^{(j)} + g_j = 0, \quad j = 1, 2,$$

where

(5.42) $$g_1 = \left(\nu\partial_{y_1}^2\tilde{W}^{(0)} + \tilde{A}_0\partial_{y_1}\tilde{W}^{(0)} + \tilde{B}_0\tilde{W}^{(0)} - i\tau\partial_{y_2}\tilde{W}^{(0)}\right) + \varkappa\tilde{W}^{(0)},$$

$$g_2 = \left(\tilde{A}_2\tilde{W}^{(0)} + \tilde{A}_1\tilde{W}^{(1)}\right) + \left(2\tilde{A}_1\partial_{y_1}\tilde{W}^{(0)} + \tilde{B}_1\partial_{y_1}\tilde{W}^{(0)} - i\frac{1}{\nu}\partial_{y_1}\tilde{W}^{(0)}\right)$$

(5.43) $$+ \left(\nu\partial_{y_1}^2\tilde{W}^{(1)} + \tilde{A}_0\partial_{y_1}\tilde{W}^{(1)} + \tilde{B}_0\tilde{W}^{(1)} - i\tau\partial_{y_2}\tilde{W}^{(1)}\right) + \varkappa\tilde{W}^{(1)}.$$

By the equivariant property, the function $\tilde{A}_0$ is odd in y_1 and $\Pi_1\tilde{A}_0 = \tilde{A}_0$. Therefore, the general solutions of homogeneous equation (5.40) has the form

(5.44) $$\tilde{W}^{(0)}(Y) = C(y_2)a^-(Y), \text{ where } a^\pm(Y) = \exp(\pm\frac{1}{2\nu}\mathfrak{D}_1^{-1}\tilde{A}_0)$$

and $C(y_2)$ is an arbitrary function. On the other hand, inhomogeneous equation (5.41) has a periodic solution if and only if

$$\text{(5.45)} \qquad \int_{-\pi}^{\pi} g_j(y_1, y_2) a^+(y_1, y_2)\, dy_1 = 0 \text{ for all } y_2.$$

Now we aim to show that solvability condition (5.45) is fulfilled for an appropriate choice of C. Substituting (5.44) into the expression for g_1 and next to (5.45) we obtain the ordinary differential equation for C,

$$-i2\pi C' + 2\pi C \varkappa - iC\tau \int_{-\pi}^{\pi} a^+ \partial_{y_2} a^- \, dy_1 + C\nu \int_{-\pi}^{\pi} a^+ \partial_{y_1}^2 a^- \, dy_1 +$$

$$+ C \int_{-\pi}^{\pi} \tilde{A}_0 a^+ \partial_{y_1} a^- \, dy_1 + C \int_{-\pi}^{\pi} \tilde{B}_0 \, dy_1 = 0.$$

Noting that

$$\int_{-\pi}^{\pi} a^+ \partial_{y_2} a^- \, dy_1 = -\frac{1}{2\nu} \int_{-\pi}^{\pi} \partial_{y_2} \mathfrak{D}_1^{-1} \tilde{A}_0 \, dy_1 = 0,$$

$$\nu \int_{-\pi}^{\pi} a^+ \partial_{y_1}^2 a^- \, dy_1 + \int_{-\pi}^{\pi} \tilde{A}_0 a^+ \partial_{y_1} a^- \, dy_1 = -\frac{1}{4\nu} \int_{-\pi}^{\pi} \tilde{A}_0^2 \, dy_1,$$

we can rewrite it in the form

$$\text{(5.46)} \qquad C'(y_2) + b(y_2) C(y_2) = 0,$$

with the coefficient

$$\text{(5.47)} \qquad b(y_2) = \frac{i}{2\pi} \int_{-\pi}^{\pi} \left(\tilde{B}_0 - \frac{1}{4\nu} \tilde{A}_0^2 \right) dy_1 + i\varkappa.$$

It follows from formula (5.7) for $\varkappa$, that the mean value of b over a period is zero. Therefore, equation (5.46) has a periodic solution

$$\text{(5.48)} \qquad C = \exp(-\mathfrak{D}_2^{-1} b).$$

In this case the particular periodic solution to (5.45) is

$$\text{(5.49)} \qquad \tilde{W}^{(1)} = -\frac{1}{2\nu} a^- \mathfrak{D}_1^{-1} (g_1 a^+).$$

Next note that, by the equivariant property, the functions $\tilde{A}_0$, $\tilde{A}_2$, and $\tilde{B}_1$ are odd and the functions $\tilde{A}_1$, $\tilde{B}_0$ are even in y_1. Hence $a^\pm$, $\tilde{W}^{(0)}$ are even, and $\tilde{W}^{(1)}$ is odd in y_1. In particular, g_2 is odd in y_1 and automatically satisfies

5.1. DESCENT METHOD

solvability condition (5.45). From this we conclude that the operators $\mathfrak{I}_j$, $j \leq 2$ vanish when

$$(5.50) \quad \tilde{W}^{(0)} = \exp\left(-\mathfrak{D}_2^{-1}b\right)a^-, \quad \tilde{W}^{(j)} = -\frac{1}{2\nu}a^-\mathfrak{D}_1^{-1}\left(g_j a^+\right), \quad j = 1, 2,$$

which along with (5.37) leads to identity (5.12).

Third step

It remains to show that operators introduced above are well-defined and satisfy inequalities (5.13). We begin with the estimating of the functions $\tilde{W}^{(p)}$. Recall that for all smooth functions $a, b \in C^s(\mathbb{R}^2/\Gamma)$,

$$\|1 - \exp a\|_{C^s} \leq c(\|a\|_{C^0})\|a\|_{C^s}, \quad \|ab\|_{C^s} \leq c(s)\|a\|_{C^0}\|b\|_{C^s} + \|a\|_{C^s}\|b\|_{C^0}.$$

It follows from this and (5.7), (5.44), (5.38), and (5.48) that

$$(5.51) \quad \|1 - \tilde{W}^{(0)}\|_{C^s} \leq c(\|\tilde{A}_0\|_{C^0}, \|\tilde{B}_0\|_{C^0})\left(\|\tilde{A}_0\|_{C^s} + \|\tilde{B}_0\|_{C^s}\right).$$

On the other hand, (5.42) implies the estimate

$$\|g_1\|_{C^s} \leq c\Big(\|1 - \tilde{W}^{(0)}\|_{C^{s+2}} + \|\tilde{A}_0\|_{C^0}\|\tilde{W}^{(0)}\|_{C^{s+1}} +$$
$$\|\tilde{A}_0\|_{C^s}\|\tilde{W}^{(0)}\|_{C^1} + \|\tilde{A}_0\|_{C^s}\Big) + \Big(\|\tilde{B}_0\|_{C^s}\|\tilde{W}^{(0)}\|_{C^0} + \|\tilde{B}_0\|_{C^0}\|\tilde{W}^{(0)}\|_{C^s}\Big).$$

Combining it with (5.51) and noting that

$$\|\tilde{W}^{(0)}\|_{C^1} \leq c\|\tilde{W}^{(0)}\|_{C^0} + c\|\tilde{W}^{(0)}\|_{C^s}$$

we arrive at

$$(5.52) \quad \|\tilde{W}^1\|_{C^s} \leq c(\|\tilde{A}_0\|_{C^0}, \|\tilde{B}_0\|_{C^0})\left(\|\tilde{A}_0\|_{C^{s+2}} + \|\tilde{B}_0\|_{C^{s+2}}\right).$$

Arguing as before and using (5.43) we obtain

$$\|g_2\|_{C^s} \leq c(\|\tilde{A}_0\|_{C^0}, \|\tilde{B}_0\|_{C^0}, \|\tilde{A}_1\|_{C^0}, \|\tilde{A}_2\|_{C^0}, \|\tilde{B}_1\|_{C^0})\Big(\|\tilde{A}_0\|_{C^{s+4}} +$$
$$\|\tilde{B}_0\|_{C^{s+4}} + \|\tilde{A}_1\|_{C^{s+3}} + \|\tilde{B}_1\|_{C^{s+2}} + \|\tilde{A}_2\|_{C^s}\Big),$$

which along with (5.50) leads to the estimate

$$(5.53) \quad \|\tilde{W}^{(2)}\|_{C^s} \leq c(\|\tilde{A}_j\|_{C^0}, \|\tilde{B}_j\|_{C^0})\Big(\sum_{j=0}^{2}\|\tilde{A}_j\|_{C^{s+4}} + \sum_{j=0}^{1}\|\tilde{B}_j\|_{C^{s+4}}\Big).$$

Noting that $\|\tilde{A}_j\|_{C^s} \leq |\mathfrak{A}|_{3,s}$, we conclude from (5.51), (5.52), (5.53) that

$$\|1 - \tilde{W}^{(0)}\|_{C^s} + \|\tilde{W}^{(1)}\|_{C^s} + \|\tilde{W}^{(2)}\|_{C^s} \leq c(|\mathfrak{A}|_{3,0}, |\mathfrak{B}|_{3,0})\Big(|\mathfrak{A}|_{3,s+4} + |\mathfrak{B}|_{3,s+4}\Big),$$

which leads to
$$(5.54)$$
$$|1 - \mathfrak{W}^{(0)}|_{m,s} + |\mathfrak{W}^{(1)}|_{m,s} + |\mathfrak{W}^{(2)}|_{m,s} \leq c(|\mathfrak{A}|_{3,0}, |\mathfrak{B}|_{3,0})\Big(|\mathfrak{A}|_{3,s+4} + |\mathfrak{B}|_{3,s+4}\Big)$$

for all $m \geq 0$. Now we can estimate the norm of the operator $\mathfrak{C}$. Since $\mathfrak{C}\Pi_1 = \mathfrak{C}$, Proposition F.1 along with (5.54) implies
(5.55)
$$\|\mathfrak{C}u\|_s \leq \|(1-\mathfrak{W}^{(0)})u\|_s + \|\mathfrak{W}^{(1)}u\|_s + \|\mathfrak{W}^{(2)}u\|_s \leq$$
$$c(|\mathfrak{A}|_{3,0}, |\mathfrak{B}|_{3,0})\Big(\big(|\mathfrak{A}|_{3,s+4} + |\mathfrak{B}|_{3,s+4}\big)\|u\|_0 + \big(|\mathfrak{A}|_{3,4} + |\mathfrak{B}|_{3,4}\big)\|u\|_s\Big).$$

Let us estimate the operators $\mathfrak{E}, \mathfrak{F}$. First note that inequalities (5.20) (5.21) from Proposition 5.4 imply the estimate

$$\|\mathfrak{U}_{S^{(p)}}u\|_{s-1} + \|\mathfrak{V}_{S^{(p)}}u\|_s \leq c(|\mathfrak{A}|_{3,0}, |\mathfrak{B}|_{3,0})\Big(\big(|\mathfrak{A}|_{3,3}|\mathfrak{W}^{(p)}|_{3,s+3}+$$
$$+|\mathfrak{A}|_{3,s+3}|\mathfrak{W}^{(p)}|_{3,3}\big)\|u\|_0 + |\mathfrak{A}|_{3,3}|\mathfrak{W}^{(p)}|_{3,3}\|u\|_{s-1}\Big).$$

From this and (5.54) we conclude that
(5.56)
$$\|\mathfrak{U}_{S^{(p)}}u\|_{s-1} + \|\mathfrak{V}_{S^{(p)}}u\|_s \leq$$
$$c(|\mathfrak{A}|_{3,7}, |\mathfrak{B}|_{3,7})\Big(\big(|\mathfrak{A}|_{3,s+7} + |\mathfrak{B}|_{3,s+7}\big)\|u\|_0 + \big(|\mathfrak{A}|_{3,7} + |\mathfrak{B}|_{3,7}\big)\|u\|_{s-1}\Big).$$

Next (5.54) along with inequality (5.24) from Lemma 5.5 implies the inequality

$$\|\mathfrak{R}^{(p)}u\|_s \leq c(|\mathfrak{A}|_{3,10}, |\mathfrak{B}|_{3,10})\big((|\mathfrak{A}|_{3,s+10} + |\mathfrak{B}|_{3,s+10})\|u\|_0 +$$
(5.57)
$$+(|\mathfrak{A}|_{3,10} + |\mathfrak{B}|_{3,10})\|u\|_{s-1}\big).$$

Recalling that for arbitrary zero-order operators $\mathfrak{A}, \mathfrak{W}$ and the operator $\mathfrak{S}$ with a symbol $S = AW$,

$$|\mathfrak{S}|_{m,s} \leq c(s,m)(|\mathfrak{A}|_{m,0}|\mathfrak{W}|_{m,s} + |\mathfrak{A}|_{m,s}|\mathfrak{W}|_{m,0}),$$

and using formula (5.23) for symbol $\mathfrak{T}$ we obtain

$$|\mathfrak{T}^{(p)}|_{3,s} \leq |1-\mathfrak{W}^{(p)}|_{3,s+2}+(|\mathfrak{A}|_{4,0}+|\mathfrak{B}|_{3,0})|\mathfrak{W}^{(p)}|_{0,s+1}+(|\mathfrak{A}|_{4,s}+|\mathfrak{B}|_{3,s})|\mathfrak{W}^{(p)}|_{0,1},$$

which being combined with (5.54) gives
(5.58) $$|\mathfrak{T}^{(p)}|_{3,s} \leq c(|\mathfrak{A}|_{4,0}, |\mathfrak{B}|_{3,0})\Big(|\mathfrak{A}|_{4,s+6} + |\mathfrak{B}|_{4,s+6}\Big).$$

In particular, inequality (5.58) along with (5.20) yields the estimate
(5.59)
$$\|\mathfrak{M}_{T^{(p)}}u\|_s + \|\mathfrak{N}_{T^{(p)}}u\|_s \leq$$
$$c(|\mathfrak{A}|_{4,0}, |\mathfrak{B}|_{4,0})\Big(\big(|\mathfrak{A}|_{4,s+6} + |\mathfrak{B}|_{4,s+6}\big)\|u\|_0 + \big(|\mathfrak{A}|_{4,9} + |\mathfrak{B}|_{4,9}\big)\|u\|_{s-1}\Big).$$

Similar arguments applying to the formula (5.23) for the symbol of the operator $\mathfrak{S}$ give the estimate
(5.60) $$|\mathfrak{S}^{(p)}|_{3,s} \leq c(|\mathfrak{A}|_{3,0}|\mathfrak{B}|_{3,0})\Big(|\mathfrak{A}|_{3,s+4} + |\mathfrak{B}|_{3,s+4}\Big).$$

Finally estimate $\mathfrak{J}_3(-\Delta)^{-1/2}$. Applying (5.54), (5.58), (5.60) to (5.35) we arrive at

$$|\mathfrak{J}_3|_{3,s} \leq c(|\mathfrak{A}|_{4,0}, |\mathfrak{B}|_{3,0})(|\mathfrak{A}|_{4,s+6} + |\mathfrak{B}|_{4,s+6}),$$

which along with inequality (F.1) from Proposition F.1 implies
(5.61)
$$\|\mathfrak{I}_3(-\Delta)^{-1/2}u\|_s \leq$$
$$c(|\mathfrak{A}|_{4,0},|\mathfrak{B}|_{3,0})\Big((|\mathfrak{A}|_{4,s+9}+|\mathfrak{B}|_{4,s+9})\|u\|_0 + (|\mathfrak{A}|_{4,9}+|\mathfrak{B}|_{4,9})\|u\|_{s-1}\Big).$$

Next note that
$$\|\mathfrak{E}u\|_s \leq \|\mathfrak{C}u\|_s + \|\mathfrak{I}_3(-\Delta)^{-1/2}u\|_s + \|\mathfrak{R}'u\|_s,$$
$$\|\mathfrak{F}u\|_s \leq \|\mathfrak{I}_3(-\Delta)^{-1/2}u\|_s + \|\mathfrak{R}''u\|_s.$$

It follows from (5.56), (5.57), and (5.59) that
$$\|\mathfrak{R}'u\|_s \leq \sum_{p=0}^{2}\|\mathfrak{U}_{S(p)}u\|_s + \sum_{p=0}^{1}\|\mathfrak{M}_{T(p)}u\|_s \leq$$
$$c(|\mathfrak{A}|_{4,9},|\mathfrak{B}|_{4,9})\Big((|\mathfrak{A}|_{4,9+s}+|\mathfrak{B}|_{4,9+s})\|u\|_0 + (|\mathfrak{A}|_{4,9}+|\mathfrak{B}|_{4,9})\|u\|_s\Big),$$
$$\|\mathfrak{R}''u\|_s \leq \sum_{p=0}^{2}\|\mathfrak{V}_{S(p)}u\|_s + \sum_{p=0}^{1}\|\mathfrak{N}_{T(p)}u\|_s + \|\mathfrak{R}^{(p)}u\|_s \leq$$
$$c(|\mathfrak{A}|_{4,10},|\mathfrak{B}|_{4,10})\Big((|\mathfrak{A}|_{4,10+s}+|\mathfrak{B}|_{4,10+s})\|u\|_0 + (|\mathfrak{A}|_{4,10}+|\mathfrak{B}|_{4,10})\|u\|_s\Big).$$

Combining these results with (5.61) and (5.13) we finally obtain
$$\|\mathfrak{C}u\|_s + \|\mathfrak{E}u\|_s \leq c(|\mathfrak{A}|_{4,9},|\mathfrak{B}|_{4,9})\Big((|\mathfrak{A}|_{4,9}+|\mathfrak{B}|_{4,9})\|u\|_s +$$
$$(|\mathfrak{A}|_{4,s+9}+|\mathfrak{B}|_{4,s+9})\|u\|_0\Big),$$
$$\|\mathfrak{F}u\|_s \leq c(|\mathfrak{A}|_{4,10},|\mathfrak{B}|_{4,10})\Big((|\mathfrak{A}|_{4,10}+|\mathfrak{B}|_{4,10})\|u\|_{s-1} +$$
$$(|\mathfrak{A}|_{4,s+10}+|\mathfrak{B}|_{4,s+10})\|u\|_0\Big),$$

which gives (5.13) and the theorem 5.2 follows.

5.2. Proof of Theorem 5.1

The proof is based on the following lemma on the invertibility of the operator $1 + \mathfrak{E}$ from Theorem 5.2.

LEMMA 5.6. *Under the assumptions of Theorem 5.1, there is a positive ε_1 depending on r and l only, so that for all $\varepsilon \in (0,\varepsilon_1)$, the operator $1 + \mathfrak{E}$ has the bounded inverse $(1+\mathfrak{E})^{-1} : H^s_{o,e}(\mathbb{R}^2/\Gamma) \mapsto H^s_{o,e}(\mathbb{R}^2/\Gamma)$ and, for all $u \in H^s_{o,e}(\mathbb{R}^2/\Gamma)$,*

(5.62) $$\|(1+\mathfrak{E})^{-1}u\|_r \leq c\|u\|_r,$$
(5.63) $$\|(1+\mathfrak{E})^{-1}u\|_s \leq c\|u\|_r(C_s + |\mathfrak{A}|_{4,s+10} + |\mathfrak{B}|_{4,s+10}) + c\|u\|_s.$$

PROOF. Formally we have
$$(1+\mathfrak{E})^{-1} = \sum_{n=0}^{\infty}(-1)^n \mathfrak{E}^n.$$

By Theorem 5.2, the operator $\mathfrak{E}: H^r_{o,e}(\mathbb{R}^2/\Gamma) \mapsto H^r_{o,e}(\mathbb{R}^2/\Gamma)$ is bounded and its norm does not exceed $c\varepsilon$. It follows from this and inequality (5.13) that for $c\varepsilon \leq 1$,

$$\|\mathfrak{E}^n u\|_s \leq c\varepsilon \|\mathfrak{E}^{n-1}u\|_s + c(|\mathfrak{A}|_{4,s+10} + |\mathfrak{B}|_{4,s+10})\|\mathfrak{E}^{n-1}u\|_r \leq$$
$$c\varepsilon\|\mathfrak{E}^{n-1}u\|_s + (c\varepsilon)^{n-1}(|\mathfrak{A}|_{4,s+10} + |\mathfrak{B}|_{4,s+10})\|u\|_r \leq$$
$$(c\varepsilon)^2\|\mathfrak{E}^{n-2}u\|_s + 2(c\varepsilon)^{n-1}(|\mathfrak{A}|_{4,s+10} + |\mathfrak{B}|_{4,s+10})\|u\|_r \leq$$
$$\ldots (c\varepsilon)^{n-1}\Big(\|u\|_s + n(|\mathfrak{A}|_{4,s+10} + |\mathfrak{B}|_{4,s+10})\|u\|_r\Big),$$

which leads to the estimate

$$\left\|\sum_{n=0}^{\infty}(-1)^n \mathfrak{E}^n u\right\|_s \leq c\Big(\|u\|_s + (|\mathfrak{A}|_{4,s+10} + |\mathfrak{B}|_{4,s+10})\|u\|_r\Big)\Big(1 + \sum_{n=1}^{\infty} n(c\varepsilon)^{n-1}\Big).$$

It remains to note that for $\varepsilon < 1/c$, the series in the right hand side converges absolutely and the lemma follows. □

Let us turn to the proof of Theorem 5.1. Since $\mathfrak{A}$ and $\mathfrak{B}$ satisfy all hypotheses of Theorem 5.2, the corresponding operators $\mathfrak{C}$, $\mathfrak{E}$, $\mathfrak{F}$ are well defined and meet all requirements of this theorem. Moreover, condition (iv) along with inequalities (5.13) yields the estimates

(5.64)
$$\|\mathfrak{C}u\|_s + \|\mathfrak{E}u\|_s \leq c\varepsilon\|u\|_s + c(|\mathfrak{A}|_{4,s+10} + |\mathfrak{B}|_{4,s+10})\|u\|_0,$$
$$\|\mathfrak{F}u\|_s \leq c\varepsilon\|u\|_{s-1} + c(|\mathfrak{A}|_{4,s+10} + |\mathfrak{B}|_{4,s+10})\|u\|_0,$$

(5.65) $\qquad \|\mathfrak{C}u\|_t + \|\mathfrak{E}u\|_t \leq c\varepsilon\|u\|_t, \quad$ for $t \in [2,r]$.

We look for a solution to basic equation (5.1) in the form

(5.66) $\qquad u = \lambda\psi^{(0)} + (1+\mathfrak{C})v$ with $v \in H^{s-1,\perp}_{o.e}$.

Substituting this representation into (5.1) we obtain the equation

(5.67) $\qquad \lambda(\mathfrak{L}+\mathfrak{H})\psi^{(0)} + (\mathfrak{L}+\mathfrak{H})(1+\mathfrak{C})v = f.$

Since Π_1 coincides with the identity mapping on $H^r_{o,e}(\mathbb{R}^2/\Gamma)$, it follows from Theorem 5.2 that

(5.68) $\qquad (\mathfrak{L}+\mathfrak{H})(1+\mathfrak{C}) = (1+\mathfrak{E})(\mathfrak{L}-\varkappa) + \mathfrak{Z}$ on $H^r_{o,e}(\mathbb{R}^2/\Gamma),$

where
$$\mathfrak{Z} = \mathfrak{F} + \mathfrak{L}_{-1}(1+\mathfrak{C}).$$

Hence we can rewrite (5.67) in the equivalent form

$$\lambda(\mathfrak{L}+\mathfrak{H})\psi^{(0)} + (1+\mathfrak{E})(\mathfrak{L}-\varkappa)v + \mathfrak{Z}v = f.$$

5.2. PROOF OF THEOREM 5.1

Applying to both sides the operators Q and $1-Q$ and noting that $Q\mathfrak{L}\psi^{(0)} = 0$ we obtain the system of operator equations for a scalar λ and unknown function $v \in H_{o,e}^{s-1,\perp}$,

$$Q(1+\mathfrak{E})(\mathfrak{L}-\varkappa)v + Q\mathfrak{Z}v = Q(f - \lambda\mathfrak{H}\psi^{(0)}), \tag{5.69}$$

$$\lambda(1-Q)(L+\mathfrak{H})\psi^{(0)} + (1-Q)(\mathfrak{L}+\mathfrak{H})(1+\mathfrak{E})v = (1-Q)f. \tag{5.70}$$

Our aim is to resolve the first equation with respect to v. Hence the task now is to prove the solvability of the equation

$$Q(1+\mathfrak{E})(\mathfrak{L}-\varkappa)v + Q\mathfrak{Z}v = Qg. \tag{5.71}$$

The corresponding result is given by

LEMMA 5.7. *Under the above assumptions, there is a positive ε_2 depending on r,l only, so that for $g \in H_{o,e}^s(\mathbb{R}^2/\Gamma)$ and $0 < \varepsilon < \varepsilon_2$, equation (5.71) has the unique solution satisfying the inequalities*

$$\|v\|_{r-1} \leq c\|g\|_r, \tag{5.72}$$

$$\|v\|_{s-1} \leq c\|g\|_r\big(C_s + |\mathfrak{A}|_{4,s+10} + |\mathfrak{B}|_{4,s+10}\big) + c\|g\|_s. \tag{5.73}$$

PROOF. We look for a solution to equation (5.71) in the form

$$v = Q(\mathfrak{L}-\varkappa)^{-1}(1+\mathfrak{E})^{-1}Q\varphi \tag{5.74}$$

with the new unknown function $\varphi \in H_{o,e}^{s,\perp}$. Since the operator $\mathfrak{L}$ commutes with the projector Q, we have

$$Q(1+\mathfrak{E})(\mathfrak{L}-\varkappa)Q(\mathfrak{L}-\varkappa)^{-1}(1+\mathfrak{E})^{-1}Q = Q\mathfrak{X}_0 Q + Q \tag{5.75}$$

with

$$\mathfrak{X}_0 = -(\mathfrak{E}Q)^2 + (1+\mathfrak{E})Q\Big((1+\mathfrak{E})^{-1} - 1 + \mathfrak{E}\Big).$$

Hence (5.71) is equivalent to the equation

$$\varphi + Q\mathfrak{X}Q\varphi = Qg, \tag{5.76}$$

in which the operator $\mathfrak{X}$ is defined by

$$\mathfrak{X} = \mathfrak{X}_0 + \mathfrak{Z}Q(\mathfrak{L}-\varkappa)^{-1}(1+\mathfrak{E})^{-1}.$$

We use the regularization method to prove the existence and uniqueness of solutions to (5.76), and consider a family of regularising equations depending on small positive parameter δ,

$$\varphi + Q\mathfrak{X}Q(-\Delta)^{-\delta}\varphi = Qg, \quad \delta > 0. \tag{5.77}$$

Let us estimate the norm of the operator $\mathfrak{X}$. It is easy to see that Lemma 5.6 implies the estimates

$$\|\mathfrak{X}_0\varphi\|_r \leq c\varepsilon^2\|\varphi\|_r, \tag{5.78}$$

$$\|\mathfrak{X}_0\varphi\|_s \leq c\|\varphi\|_r\big(|\mathfrak{A}|_{4,s+10} + |\mathfrak{B}|_{4,s+10}\big) + c\varepsilon^2\|\varphi\|_s.$$

On the other hand, inequality (5.8) along with Lemma 5.6 leads to the estimate

(5.79) $$\|\mathcal{Q}(\mathfrak{L} - \varkappa)^{-1}(1+\mathfrak{E})^{-1}\mathcal{Q}\varphi\|_{r-1} \leq c\|\varphi\|_r,$$
$$\|\mathcal{Q}(\mathfrak{L} - \varkappa)^{-1}(1+\mathfrak{E})^{-1}\mathcal{Q}\varphi\|_{s-1} \leq c\|\varphi\|_r\big(|\mathfrak{A}|_{4,s+10} + |\mathfrak{B}|_{4,s+10}\big) + c\|\varphi\|_s.$$

Next estimates (5.64) for $\mathfrak{C}$ and $\mathfrak{F}$ along with inequalities (5.6) for $\mathfrak{L}_{-1}$ imply

(5.80) $$\|\mathfrak{Z}\varphi\|_r \leq c\varepsilon\|\varphi\|_{r-1},$$
$$\|\mathfrak{Z}\varphi\|_s \leq c\|\varphi\|_{r-1}\big(C_s + |\mathfrak{A}|_{4,s+10} + |\mathfrak{B}|_{4,s+10}\big) + c\varepsilon\|\varphi\|_{s-1}.$$

Combining (5.78)-(5.80) we finally arrive at

$$\|\mathfrak{X}\varphi\|_r \leq c\varepsilon\|\varphi\|_r,$$
$$\|\mathfrak{X}\varphi\|_s \leq c\|\varphi\|_r\big(C_s + |\mathfrak{A}|_{4,s+10} + |\mathfrak{B}|_{4,s+10}\big) + c\varepsilon\|\varphi\|_s.$$

Since the operators $(-\Delta)^{-\delta}$ are uniformly bounded in $H^r_{o,e}(\mathbb{R}^2/\Gamma)$, it follows that each solution to equation (5.77) satisfies the inequalities

$$\|\varphi\|_r(1 - c\varepsilon) \leq c\|g\|_r,$$
$$\|\varphi\|_s(1 - c\varepsilon) \leq c\|\varphi\|_r\big(C_s + |\mathfrak{A}|_{4,s+10} + |\mathfrak{B}|_{4,s+10}\big) +$$
$$c\|g\|_s + c\|g\|_r\big(|\mathfrak{A}|_{4,s+10} + |\mathfrak{B}|_{4,s+10}\big).$$

Hence for $\varepsilon < 1/2c$,

(5.81) $$\|\varphi\|_r \leq c\|g\|_r,$$
$$\|\varphi\|_s \leq c\|g\|_s + c\|g\|_r\big(C_s + |\mathfrak{A}|_{4,s+10} + |\mathfrak{B}|_{4,s+10}\big),$$

In particular, the solution is unique. Since the operator

$$\mathcal{Q}\mathfrak{X}\mathcal{Q}(-\Delta)^{-\delta} : H^r_{o,e}(\mathbb{R}^2/\Gamma) \mapsto H^r_{o,e}(\mathbb{R}^2/\Gamma)$$

is compact, uniqueness along with the Fredholm Theorem implies the solvability of equations (5.77) for all $\delta > 0$. Hence they have a family of solutions φ_δ satisfying (5.81). After passing to a subsequence we can assume that

$$\varphi_\delta \to \varphi, \quad (-\Delta)^{-\delta}\varphi_\delta \to \varphi \text{ weakly in } H^s(\mathbb{R}^2/\Gamma) \text{ as } \delta \searrow 0.$$

Obviously φ serves as a solution to equation (5.76). Recalling (5.74) we obtain the solvability and uniqueness result for (5.71). It remains to note that estimates (5.72), (5.73) follow from formula (5.74) and inequalities (5.79),(5.81). $\square$

We are now in a position to complete the proof of Theorem 5.1. Applying Lemma 5.7 to equation (5.69) we obtain the representation

(5.82) $$v = \lambda v_0 + v_f,$$

in which v_0 and v_f are solutions to equations (5.71) with $g = -\mathcal{Q}\mathfrak{H}\psi^{(0)}$ and $g = \mathcal{Q}f$. It follows from (5.72), (5.73) that

$$
(5.83) \quad \begin{aligned}
&\|v_f\|_{r-1} \leq c\|f\|_r, \\
&\|v_f\|_{s-1} \leq c\|f\|_r \big(C_s + |\mathfrak{A}|_{4,s+10} + |\mathfrak{B}|_{4,s+10}\big) + c\|f\|_s, \\
&\|v_0\|_{r-1} \leq c\varepsilon, \\
&\|v_0\|_{s-1} \leq c\big(C_s + |\mathfrak{A}|_{4,s+10} + |\mathfrak{B}|_{4,s+10}\big).
\end{aligned}
$$

The functions v_0 and v_f are completely defined by the eigenfunction ψ_0 and the right hand side f of equation (5.1), but the scalar λ still remains unknown. In order to find it we substitute representation (5.82) into (5.70) to obtain

$$(5.84) \quad \lambda(1-\mathcal{Q})(\mathfrak{L}+\mathfrak{H})\Big(\psi^{(0)} + (1+\mathfrak{C})v_0\Big) = (1-\mathcal{Q})\Big(f - (\mathfrak{L}+\mathfrak{H})(1+\mathfrak{C})v_f\Big).$$

It follows from identity (5.68) that equation (5.71) for v_0 is equivalent to

$$\mathcal{Q}(\mathfrak{L}-\varkappa)\mathcal{Q}v_0 + \mathcal{Q}\big((\mathfrak{L}+\mathfrak{H})\mathfrak{C} + \mathfrak{H} + \kappa\big)\mathcal{Q}v_0 = -\mathcal{Q}\mathfrak{H}\psi^{(0)}.$$

Since the operator $\mathcal{Q}(\mathfrak{L}-\varkappa)$ has the bounded inverse $\mathcal{Q}(\mathfrak{L}-\varkappa)^{-1} : H^{s,\perp}_{o,e} \mapsto H^{s-1,\perp}_{o,e}$, we have

$$(5.85) \quad v_0 + \mathcal{Q}(\mathfrak{L}-\varkappa)^{-1}\mathcal{Q}\big((\mathfrak{L}+\mathfrak{H})\mathfrak{C} + \mathfrak{H} + \kappa\big)\mathcal{Q}v_0 = -\mathcal{Q}(\mathfrak{L}-\varkappa)^{-1}\mathcal{Q}\mathfrak{H}\psi^{(0)}.$$

Next, using inequalities (5.83) along with (5.65) and (5.72) and noting that for $2 \leq t \leq r$,

$$\|\mathfrak{H}u\|_{t-1} \leq c\varepsilon\|u\|_t$$

we arrive at

$$\|\mathcal{Q}(\mathfrak{L}-\varkappa)^{-1}\mathcal{Q}\mathfrak{H}\psi^{(0)}\|_{r-2} \leq c\varepsilon, \quad \|\mathcal{Q}(\mathfrak{L}-\varkappa)^{-1}\big((\mathfrak{L}+\mathfrak{H})\mathfrak{C}+\mathfrak{H}+\varkappa\big)v_0\|_{r-3} \leq c\varepsilon^2,$$

which leads to

$$v_0 = -\mathcal{Q}(\mathfrak{L}-\varkappa)^{-1}\mathcal{Q}\mathfrak{H}\psi^{(0)} + v' \quad \text{with} \quad \|v'\|_{r-3} \leq c\varepsilon^2.$$

Substituting this expression into (5.84) gives
(5.86)
$$\lambda(1-\mathcal{Q})(\mathfrak{L}+\mathfrak{H})\Big(\psi^{(0)} - \mathcal{Q}(\mathfrak{L}-\varkappa)^{-1}\mathcal{Q}\mathfrak{H}\psi^{(0)}\Big) + \lambda v'' = (1-\mathcal{Q})\Big(f - (\mathfrak{L}+\mathfrak{H})(1+\mathfrak{C})v_f\Big)$$

where

$$v'' = (1-\mathcal{Q})(\mathfrak{L}+\mathfrak{H})\Big(v' + \mathfrak{C}v_0\Big).$$

Next noting that

$$\|(1-\mathcal{Q})\mathfrak{L}u\|_s \leq c\varepsilon\|u\|_0$$

and using estimate (5.65) we obtain

$$\|v''\|_{r-4} \leq c\varepsilon\|v'\|_{r-3} + c\varepsilon^2\|v_0\|_{r-2} \leq c\varepsilon^3.$$

Multiplying both sides of (5.84) by $\psi^{(0)}$ and integrating the result over $\mathbb{T}^2$ leads to equality

$$\lambda K = \int_{\mathbb{T}^2} \Big(f - (\mathfrak{L} + \mathfrak{H})(1 + \mathfrak{C})v_f\Big)\psi^{(0)}\,dY, \tag{5.87}$$

in which

$$K = \int_{\mathbb{T}^2}(\mathfrak{L}+\mathfrak{H})\Big(\psi^{(0)} - \mathcal{Q}(\mathfrak{L}-\varkappa)^{-1}\mathcal{Q}\mathfrak{H}\psi^{(0)}\Big)\psi^{(0)}\,dY + \int_{\mathbb{T}^2} v''\psi^{(0)}\,dY = @\varepsilon^2 + O(\varepsilon^3).$$

Hence there is a positive $\varepsilon_1 < \varepsilon_2$ depending on r,l and ψ_0 only so that $2|K| > |@|\varepsilon^2$ for all $0 < \varepsilon < \varepsilon_1$. From this, relation (5.87) and inequalities (5.83) we conclude that for such ε the unknown λ is well defined and

$$\varepsilon^2|\lambda| < \frac{c}{|@|}\|f\|_r. \tag{5.88}$$

Hence for $0 < \varepsilon < \varepsilon_1$, equation (5.1) has the unique solution $u \in H^r_{o,e}$. It remains to note that inequalities (5.83) and (5.88) along with the identity

$$u = \lambda\psi^{(0)} + \lambda(1 + \mathfrak{C})v_0 + (1 + \mathfrak{C})v_f$$

imply estimates (5.10), (5.11) and this ends the proof of Theorem 5.1.

5.3. Verification of assumptions of Theorem 5.1

In this section we check all abstract conditions required for solving the linear equation (5.1), in the case when the operators involved in this equation are defined by Theorem 3.4. Note that the symmetry conditions (5.3) and (iii) follows from assertion (iii) of this theorem. The same conclusion can be drawn for the Metric conditions (5.5), (5.6) which are realized, as soon as $\|U\|_\rho \leq \varepsilon$ and $\rho = l + 3$.

Let us consider the restrictions on the spectrum and resolvent of $\mathfrak{L}$. Part (v) results immediately from Theorem 4.19, restricted to the space of functions odd in y_1 and even in y_2, and once we notice that the eigenvalue corresponding to the eigenvector $\psi^{(0)} = \sin y_1 \cos y_2$ satisfies

$$-\nu + \sqrt{1 + \tau^2} = \varepsilon^2\nu_1(\tau) + O(\varepsilon^3). \tag{5.89}$$

This results immediately from the fact that the linear operator $\mathcal{L}$ (see (3.3) corresponds to the differentiation of the basic system at the point

$$U = U_\varepsilon^{(N)} + \varepsilon^N W, \quad N \geq 3 \tag{5.90}$$

where $U_\varepsilon^{(N)}$ is the approximate solution at order ε^N, and from Lemma 3.7. Moreover the coefficient ν_1 is positive (larger than a positive number) for any value of τ.

Now, from Lemma 3.8 we see that the Fourier expansion of the diffeomorphism of the torus has, at order ε, only harmonic 1 terms in Y, and at order ε^2 only harmonics 2 and 0 in $Y = (y_1, y_2)$, all terms being invariant under the shift $\mathcal{T}_{\mathbf{v}_0} : Y \mapsto Y + (\pi, \pi)$. Let us now consider the expression of

5.3. VERIFICATION OF ASSUMPTIONS OF THEOREM 5.1

the linear operator $\mathfrak{L} + \mathfrak{H}$ which is obtained after applying the above diffeomorphism on a linear operator which may be formally expanded in powers of ε, having coefficients with the same property as the diffeomorphism in terms of harmonics in Y. The result is that the formal expansion in powers of ε of $\mathfrak{L} + \mathfrak{H}$ has the same property as above i.e. the order 0 is independent of Y, while orders ε and ε^2 only contain respectively harmonic 1 and harmonics 2 and 0 in Y.

Let us consider the formula (5.7) giving the coefficient $\varkappa$. In the integral over $\mathbb{T}^2$ the functions A and B are $O(\varepsilon)$ and can be expanded in powers of ε, their principal part containing only harmonic 1-terms in Y. It results immediately that

$$(5.91) \qquad \varkappa = O(\varepsilon^2).$$

It is not useful in our proof to give more precision on this coefficient, however the interested reader might check (after few days of computations) that

$$(5.92) \qquad \varkappa = \varepsilon^2 \frac{\mu_c(\tau)}{16} \{8\tau^2 - 5\tau^2 - \frac{7}{4}\} + O(\varepsilon^3).$$

Now, from the estimate on operators $\mathfrak{A}$ and $\mathfrak{B}$ in theorem 3.4 and from (5.7) we have

$$|\varkappa| \leq c \|U\|_{14},$$

hence from (5.91) and (5.90) we deduce that

$$(5.93) \qquad \varkappa = \varepsilon^2 \widetilde{\varkappa}, \quad |\widetilde{\varkappa}| \leq c(1 + \varepsilon \|W\|_{14}).$$

It should be noticed that in formulae (5.89) and (5.92) the terms of order $O(\varepsilon^3)$ depend on the unknown W of (5.90) which is assumed to be bounded in $\mathbb{H}^{14}_{(S)}$ (because of the smoothness required for computing operators $\mathfrak{B}$ and $\mathfrak{L}_{-1}$ above). Hence, if we want to apply the result of Theorem 4.2 for obtaining the required estimate (5.8) for $(\mathfrak{L} - \varkappa)^{-1}$ we need to show that properties (4.4) hold for the functions of ε

$$\nu_j(\varepsilon) = \nu_0 - \varepsilon^2 \nu_1 + \varepsilon^3 \widetilde{\nu}_j(\varepsilon), \quad \varkappa = \varepsilon^2 \widetilde{\varkappa}_j(\varepsilon)$$

where $\nu_0 = (1 + \tau^2)^{1/2}$, $\nu_1 = \nu_1(\tau)$, $\widetilde{\nu}_j(\varepsilon)$ and $\widetilde{\varkappa}_j(\varepsilon)$ are obtained with some W_j in $\mathbb{H}^{14}_{(S)}$ through the iteration of the Newton method of Nash Moser theorem. Because of the smooth dependence of these coefficients in function of (ε, U) (see in particular Theorem 3.4 for ν, and (5.93) for $\varkappa$), the properties (4.4) for $\widetilde{\nu}_j(\varepsilon)$ and $\widetilde{\varkappa}_j(\varepsilon)$ are verified as soon as there exists $R > 0$ such that

$$(5.94) \qquad \begin{aligned} \|W_j(\varepsilon') - W_j(\varepsilon'')\|_{14} &\leq R|\varepsilon' - \varepsilon''|, \\ \|W_{j+1}(\varepsilon) - W_j(\varepsilon)\|_{14} &\leq 2^{-j} R \end{aligned}$$

holds. This property will be checked in Chapter 6. Assuming that this is true, we have all conditions of Theorem 4.2 realized. This completes the verification of restrictions on the spectrum and resolvent of $\mathfrak{L}$.

The nondegeneracy condition (5.9) is fundamental for obtaining the estimates (5.10), (5.11) for the solution of the linearized system. So, let us now consider the coefficient @ defined in (5.9). We notice that the eigenfunction $\psi^{(0)} = \sin y_1 \cos y_2$ only contains harmonic 1-terms in Y, and assuming that W in formula (5.90) is bounded in $\mathbb{H}^{14}_{(S)}$, and denoting the coefficient of order ε in the operator $\mathfrak{H}$ by $\mathfrak{H}^{(1)}$ we have

$$\mathcal{Q}\mathfrak{H}\psi^{(0)} = \varepsilon\mathfrak{H}^{(1)}\psi^{(0)} + O(\varepsilon^2),$$

and, since $\mathfrak{L}_0$ is invertible on finite Fourier series orthogonal to $\psi^{(0)}$, one obtains

$$\mathfrak{H}\mathcal{Q}(\mathfrak{L} - \varkappa)^{-1}\mathcal{Q}\mathfrak{H}\psi^{(0)} = \varepsilon^2\mathfrak{H}^{(1)}\mathfrak{L}_0^{-1}\mathfrak{H}^{(1)}\psi^{(0)} + O(\varepsilon^3).$$

Now defining the coefficient of order ε^2 in the operator $\mathfrak{L} + \mathfrak{H}$ by $\mathfrak{H}^{(2)}$ we have

$$(5.95) \qquad @ = \int_{\mathbb{T}^2} \left(\mathfrak{H}^{(2)}\psi^{(0)} - \mathfrak{H}^{(1)}\mathfrak{L}_0^{-1}\mathfrak{H}^{(1)}\psi^{(0)} \right) \psi^{(0)} \, dY.$$

Then, we prove the following

LEMMA 5.8. *The coefficient @ of nondegeneracy condition (5.9) is given by*

$$@ = -2\pi^2 \frac{\mu_1}{\mu_0^2} = \frac{\pi^2}{8}(\alpha_0 + \beta_0),$$

where $\mu_1(\tau)$ is given by (2.11) and $(\alpha_0 + \beta_0)(\tau)$ is given at Theorem 2.3. This coefficient is non zero for $\tau \neq \tau_c$.

Proof: the proof is made at Appendix H.

This completes the verification of Nondegeneracy condition.

5.4. Inversion of $\mathcal{L}$

Let us now consider the inversion of the operator $\mathcal{L}$ defined in (3.3)

$$(5.96) \qquad \mathcal{L}(U,\mu)V = F,$$

where

$$\mathcal{L}(U,\mu) = \begin{pmatrix} \mathcal{G}_\eta & \mathcal{J}^* \\ \mathcal{J} & \mathfrak{a} \end{pmatrix},$$
$$F = (f, g) \in \mathbb{H}^s_{(S)}, \quad V = (\delta\phi, \delta\eta)$$

and where we look for

$$\delta U = (\delta\psi, \delta\eta), \quad \delta\psi = \delta\phi + \mathfrak{b}\delta\eta,$$

where $\mathfrak{b}$ is defined in (3.1). In this section we prove the following Theorem which collects the results of previoussections on the operator $\mathcal{L}$

THEOREM 5.9. *Consider $M > 0$, $\imath \geq 14$, and $\pi/4 > \delta > 0$, and set*

$$(5.97) \qquad U = U^{(N)}_\varepsilon + \varepsilon^N W, \quad N \geq 3,$$
$$\mu = \mu^{(N)}_\varepsilon = \mu_c + \varepsilon^2 \mu_1(\tau) + O(\varepsilon^3),$$

where $\|W\|_\imath \leq M$, and $(U^{(N)}_\varepsilon, \mu^{(N)}_\varepsilon)$ is the approximate solution at order ε^N obtained at Theorem 2.3 in the case of diamond waves ($\varepsilon_1 = \varepsilon_2 = \varepsilon/2$), and $\tau = \tan\theta_0$, $\mu_c = \cos\theta_0$, $\delta < \theta_0 < \pi/2 - \delta$. Assume moreover in (5.97) that $W = W_j(\varepsilon)$, $j \in \mathbb{N}$ satisfies the property (5.94), then for $\mu_c^{-1} \in \mathfrak{N}_\alpha$, $\alpha \in (0, 1/78)$, there exists $\varepsilon_0 > 0$ and a subset $\mathcal{E}$ of $[0, \varepsilon_0]$ such that for any $F \in \mathbb{H}_{(S)}^s$, $s \geq 5$, and for $\varepsilon \in \mathcal{E}$, the linear equation (5.96) has a unique solution V corresponding to $\delta U \in \mathbb{H}_{(S)}^{s-3}$, such that the following estimates hold

(5.98) $\quad \|\delta U\|_{r-3} \leq \dfrac{c}{\varepsilon^2}\|F\|_r, \quad 5 \leq r \leq \iota - 8,$

$\|\delta U\|_{s-3} \leq \dfrac{c(s)}{\varepsilon^2}\|F\|_r(1 + \varepsilon^N\|W_j\|_{s+18}) + c(s)\|F\|_s.$

Moreover the following property holds for the "good" set $\mathcal{E}$:

(5.99) $\quad \dfrac{1}{|\mu_r^{(N)} - \mu_c|} meas\{\mu = \mu_\varepsilon^{(N)}; \varepsilon \in \mathcal{E} \cap (0,r)\} \to 1, \quad as\ r \to 0.$

Proof. By construction, we have
$$\mathcal{G}_\eta(\delta\phi) - \mathcal{J}^*(\tfrac{1}{\mathfrak{a}}\mathcal{J}(\delta\phi)) = h,$$
where
$$h = f - \mathcal{J}^*(g/\mathfrak{a}),$$
and which, after the change of coordinates of Theorem 3.4, becomes
$$(\mathfrak{L} + \mathfrak{A}\mathfrak{D}_1 + \mathfrak{B} + \mathfrak{L}_{-1})\widetilde{\delta\phi} = \dfrac{\widetilde{h}}{\kappa},$$
which we learnt to invert at Theorem 5.1.

First, for $U \in \mathbb{H}_{(S)}^m$, and $\|U\|_r \leq M_r$, we have the following estimates, due to Lemma 3.1
$$\|h\|_s \leq c(M_4)\{\|F\|_{s+1} + \|F\|_2\|U\|_{s+2}),$$
and with Theorem 3.4 we obtain
$$\|\widetilde{h}\|_s \leq c(M_4)(\|h\|_s + \|U\|_{s+4}\|h\|_0).$$
Taking into account of the estimate on $\kappa(Y)$ in Theorem 3.4, we then arrive to

(5.100) $\quad \|\dfrac{\widetilde{h}}{\kappa}\|_s \leq c(M_5)\{\|F\|_{s+1} + \|F\|_2\|U\|_{s+5}\}.$

In the same way, thanks to the Theorem 3.4, we also have
$$\|\delta\phi\|_s \leq c(M_4)\{\|\widetilde{\delta\phi}\|_s + \|U\|_{s+4}\|\widetilde{\delta\phi}\|_0\},$$
and since
$$\delta\eta = -\dfrac{1}{\mathfrak{a}}\mathcal{J}(\delta\phi) + \dfrac{1}{\mathfrak{a}}g,$$
we obtain
$$\begin{aligned}\|\delta\eta\|_s \leq\ & c(M_4)\{\|\widetilde{\delta\phi}\|_{s+1} + \|U\|_{s+5}\|\widetilde{\delta\phi}\|_0\} + \\ & + c(M_4)\{\|F\|_s + \|U\|_{s+2}\|F\|_2\}.\end{aligned}$$

Hence
$$\|V\|_s \leq c(M_4)\{\|\widetilde{\delta\phi}\|_{s+1} + \|U\|_{s+5}\|\widetilde{\delta\phi}\|_0\} + \\ + c(M_4)\{\|F\|_s + \|U\|_{s+2}\|F\|_2\},$$

and thanks to Lemma 3.1

$$\|\delta U\|_s \leq c(M_3)\{\|V\|_s + \|U\|_{s+1}\|V\|_2\},$$
$$\leq c(M_7)\{\|\widetilde{\delta\phi}\|_{s+1} + \|U\|_{s+5}\|\widetilde{\delta\phi}\|_0 + \|U\|_{s+1}\|\widetilde{\delta\phi}\|_3\} +$$
(5.101) $$\quad + c(M_4)\{\|F\|_s + \|U\|_{s+2}\|F\|_2\}.$$

It remains to use Theorem 5.1 which connects $\widetilde{\delta\phi}$ and $\frac{\widetilde{h}}{\kappa}$, taking into account of estimates ii) of Theorem 3.4. Assuming that properties (4.4) hold for $\widetilde{\nu}_j(\varepsilon)$ and $\widetilde{\varkappa}_j(\varepsilon)$ we have (see Theorem 4.2) for $\varepsilon \leq \varepsilon_0$ and $\varepsilon \in \mathfrak{N}_\alpha$, $\alpha \in (0, 1/78)$ the following estimates for $\|U\|_\rho \leq \varepsilon$, $\imath \geq 14$, $1 \leq r \leq s$

$$\|\widetilde{\delta\phi}\|_{r-1} \leq \frac{c(M_{14})}{\varepsilon^2}\|\frac{\widetilde{h}}{\kappa}\|_r, \quad 1 \leq r \leq \imath - 9$$

$$\|\widetilde{\delta\phi}\|_{s-1} \leq \frac{c(M_{14})}{\varepsilon^2}\|\frac{\widetilde{h}}{\kappa}\|_r(1 + \|U\|_{s+19}) + c(M_{14})\|\frac{\widetilde{h}}{\kappa}\|_s.$$

From this and (5.100) we deduce that

$$\|\widetilde{\delta\phi}\|_{r-2} \leq \frac{c(M_{14})}{\varepsilon^2}\|F\|_r, \quad 2 \leq r \leq \imath - 8,$$

$$\|\widetilde{\delta\phi}\|_{s-2} \leq \frac{c(M_{14})}{\varepsilon^2}\|F\|_r(1 + \|U\|_{s+18}) + c(M_{14})\|F\|_s,$$

and from (5.101) we obtain the estimates for δU (5.98). The rest of Theorem 5.9 follows directly from the results of previous section and from Theorem 4.2.

CHAPTER 6

Nonlinear Problem. Proof of Theorem 1.3

In this chapter we complete the proof of the main Theorem 1.3 on existence of diamond nonlinear waves of finite amplitude. To this end we exploit the general version of the Nash-Moser Implicit Function Theorem proved in Appendix N of [**IPT**]. This result concerns the solvability of the operator equation

(6.1) $$\Phi(W, \varepsilon) = 0$$

in scales of Banach spaces E_s and F_s parametrized by $s \in \mathbb{N}_0 = \mathbb{N} \cup \{0\}$, and supplemented with the norms $\|\cdot\|_s$ and $|\cdot|_s$. It is supposed that they satisfy the following conditions.

(A1) For $t < s$ there exists $c(t,s)$ such that
$$\|\cdot\|_t \leq c(t,s)\|\cdot\|_s, \qquad |\cdot|_t \leq c(t,s)|\cdot|_s.$$
(A2) For $\lambda \in [0,1]$ with $\lambda t + (1-\lambda)s \in \mathbb{N}$,
$$\|\cdot\|_{\lambda t + (1-\lambda)s} \leq c(t,s)\|\cdot\|_t^\lambda \|\cdot\|_s^{1-\lambda}, \quad |\cdot|_{\lambda t + (1-\lambda)s} \leq c(t,s)|\cdot|_t^\lambda |\cdot|_s^{1-\lambda}.$$

(A3) There exists a family of smoothing operators $S_\wp$ defined over the first scale such that for $\wp > 0$ and $t < s$,
$$\|S_\wp W\|_t \leq c(t,s)\|W\|_s, \qquad \|S_\wp W\|_s \leq c(t,s)\wp^{t-s}\|W\|_t,$$
$$\|S_\wp W - W\|_t \leq c(t,s)\wp^{s-t}\|W\|_s,$$

and, if $\varepsilon \mapsto \wp(\varepsilon)$ is a smooth, increasing, convex function on $[0, \infty)$ with $\wp(0) = 0$, then, for $0 < \varepsilon_1 < \varepsilon_2$,
$$\|(S_{\wp(\varepsilon_1)} - S_{\wp(\varepsilon_2)})W\|_s \leq c(t,s)|\varepsilon_1 - \varepsilon_2|\wp'(\varepsilon_2)\wp(\varepsilon_1)^{t-s-1}\|W\|_t.$$

(B1) Operators $\Phi(\cdot, \varepsilon)$, depend on a small parameter $\varepsilon \in [0, \varepsilon_0]$, and map a neighborhood of 0 in E_r into F_ρ. Suppose that there exist
$$\sigma \leq \rho \leq r - 1, \quad \sigma, \rho, r \in \mathbb{N}_0,$$

and, for all $l \in \mathbb{N}_0$, numbers $c(l) > 0$ and $\varepsilon(l) \in (0, \varepsilon_0]$ with the following properties for all $W, U, W_i, U_i \in B$ and $\varepsilon, \varepsilon_i \in [0, \varepsilon_0]$, $i = 1, 2$, where $B = \{W \in E_r : |W|_r \leq R_0\}$ for some $R_0 > 0$:

(B2) The operator $\Phi : B \times [0, \varepsilon_0] \to F_\rho$ is twice continuously differentiable,

(6.2) $$|\Phi(W, \varepsilon)|_{\rho+l} \leq c(l)(1 + \|W\|_{r+l})$$

and, for $W, U \in E_{r+l}$, $\varepsilon \in [0, \varepsilon(l)]$,

(6.3)
$$|D(W,U,\varepsilon)|_{\rho+l} \leq c(l)(1 + \|W\|_{r+l} + \|U\|_{r+l})\|W - U\|_r^2 + \\ + c(l)\|W - U\|_r \|W - U\|_{r+l},$$

where
$$D(W, U, \varepsilon) = \Phi(W, \varepsilon) - \Phi(U, \varepsilon) - \Phi'_W(U, \varepsilon)(W - U).$$

Moreover,

(6.4)
$$|D(W_1, U_1, \varepsilon_1) - D(W_2, U_2, \varepsilon_2)|_\rho \leq c\big(|\varepsilon_1 - \varepsilon_2| + \|W_1 - W_2\|_r + \|W_1 - W_2\|_r\big) \\ \cdot \big(\|W_1 - U_1\|_r + \|W_2 - U_2\|_r\big).$$

(B3) There exists a family of bounded linear operators $\Lambda(W, \varepsilon) : E_r \to F_\rho$, depending on $(W, \varepsilon) \in B \times [0, \varepsilon_0]$, with

(6.5)
$$|\Lambda(W, \varepsilon)U|_\rho \leq c(0)\|U\|_r, \ U \in E_r,$$

that approximates the Fréchet derivative Φ'_W as follows. For $W \in E_{r+l} \cap B$, $\varepsilon \in [0, \varepsilon(l)]$ and $U \in E_{r+l}$,

(6.6)
$$|\Lambda(W, \varepsilon)U - \Phi'_W(W, \varepsilon)U|_{\rho+l} \leq c(l)(1 + \|W\|_{r+l})|\Phi(W, \varepsilon)|_r \|U\|_r + \\ + c(l)|\Phi(W, \varepsilon)|_{r+l}\|U\|_r + c(l)|\Phi(W, \varepsilon)|_r \|U\|_{r+l}.$$

(B4) When $W_i \in B \cap E_{r+l}$, $\varepsilon_i \in [0, \varepsilon(l)]$, $i = 1, 2$,

(6.7)
$$|\Phi(W_1, \varepsilon_1) - \Phi(W_2, \varepsilon_2)|_{\rho+l} \leq c(l)\big(1 + \|W_1\|_{r+l} + \|W_2\|_{r+l}\big) \\ \cdot \big(|\varepsilon_1 - \varepsilon_2| + \|W_1 - W_2\|_r\big) + c(l)\|W_1 - W_2\|_{r+l},$$

(6.8)
$$|(\Phi'_W(W_1, \varepsilon_1) - \Phi'_W(W_2, \varepsilon_2))U|_{\rho+l} + |(\Lambda(W_1, \varepsilon_1) - \Lambda(W_2, \varepsilon_2))U|_{\rho+l} \leq \\ \leq c(l)\Big(\|W_1 - W_2\|_{r+l} + (|\varepsilon_1 - \varepsilon_2| + \|W_1 - W_2\|_r)(\|W_1\|_{r+l} + \|W_2\|_{r+l})\Big)\|U\|_r \\ + \big(|\varepsilon_1 - \varepsilon_2| + \|W_1 - W_2\|_r\big)\|U\|_{r+l},$$

- A set $\mathcal{E} \subset [0, \infty)$ is dense at 0 if $\displaystyle\lim_{r \searrow 0} \frac{2}{r^2} \int_{\mathcal{E} \cap [0, r]} \varepsilon \, d\varepsilon = 1$.

(B5) If a set $\mathcal{E} \subset [0, \varepsilon(l)]$ is dense at 0 and a mapping $\vartheta : \mathcal{E} \to B \cap E_{r+l}$ is Lipschitz in the sense that for $\varepsilon_1, \varepsilon_2 \in \mathcal{E}$,

$$\|\vartheta(\varepsilon_1) - \vartheta(\varepsilon_2)\|_r \leq C|\varepsilon_1 - \varepsilon_2| \text{ where } C = C(\vartheta), \text{ constant,}$$

then there is a set $\mathcal{E}(\vartheta) \subset \mathcal{E}$, which is also dense at 0, such that, for any $\varepsilon \in \mathcal{E}(\vartheta)$ and $f \in F_{\rho+l}$, the equation $\Lambda(\vartheta(\varepsilon), \varepsilon)\delta W = f$ has a unique solution satisfying

(6.9)
$$\|\delta W\|_{\rho-\sigma+l} \leq \varepsilon^{-\varrho} c(l)(|f|_{\rho+l} + \|\vartheta(\varepsilon)\|_{r+l}|f|_\rho).$$

6. NONLINEAR PROBLEM

(B6) Suppose that $\vartheta_0 : \mathcal{E}_0 \to B \cap E_{r+l}$ and mappings $\vartheta_k : \cap_{i=0}^{k-1}\mathcal{E}(\vartheta_i) \to B \cap E_{r+l}$ satisfy, for a constant C independent of $k \subset \mathbb{N}$ sufficiently large,

$$\|\vartheta_k(\varepsilon_1) - \vartheta_k(\varepsilon_2)\|_r \leq C|\varepsilon_1 - \varepsilon_2|, \quad \varepsilon_1, \varepsilon_2 \in \cap_{j=0}^{k-1}\mathcal{E}(\vartheta_j),$$

$$\|\vartheta_{k+1}(\varepsilon) - \vartheta_k(\varepsilon)\|_r \leq \frac{1}{2^k}, \quad \varepsilon \in \cap_{j=0}^{k}\mathcal{E}(\vartheta_j).$$

Then $\cap_{j=0}^{\infty}\mathcal{E}(\vartheta_j)$ is dense at 0, where the sets $\mathcal{E}(\vartheta_j)$ are defined in (B5).

THEOREM 6.1. *Suppose (A1)–(B6) hold and, for $N \in \mathbb{N}$ with $N \geq 2$, equation (6.1) has approximate solution $W = W_\varepsilon^{(N)} \in \cap_{s \in \mathbb{N}_0} E_s$, with, for a constant $k(N,s)$,*

(6.10) $\qquad \|W_\varepsilon^{(N)}\|_s \leq k(N,s)\varepsilon, \quad |\Phi(\varepsilon, W_\varepsilon^{(N)})|_s \leq k(N,s)|\varepsilon|^{N+1}$

and

(6.11) $\qquad \|W_{\varepsilon_1}^{(N)} - W_{\varepsilon_2}^{(N)}\|_s \leq k(N,s)|\varepsilon_1 - \varepsilon_2|.$

Then there is a set $\mathcal{E}$, which is dense at 0, and a family

$$\{W = \vartheta(\varepsilon) : \varepsilon \in \mathcal{E}\} \subset E_r$$

of solutions to (6.1) with $\|\vartheta(\varepsilon_1) - \vartheta(\varepsilon_2)\|_r \leq c|\varepsilon_1 - \varepsilon_2|$ for some constant c.

In order to apply Theorem 6.1 to the 3D wave problem, let us introduce some notations. Fix an arbitrary $\alpha \in (0, 1/78)$ and $0 < \delta < 1$. Denote by $\mathcal{N}$ the set of all μ_c so that $\nu_0 = \mu_c^{-1}$ belongs to the set $\mathfrak{N}_\alpha$ given by Theorem 4.2. Since $\mathfrak{N}_\alpha \subset [1, \infty)$ is a set of full measure, $\mathfrak{N}$ is the set of full measure in $(0,1)$. Choose an arbitrary $\mu_c = (1+\tau^2)^{-1/2} \in \mathfrak{N}$ so that $\tau \in (\delta, 1/\delta)$ and set $\tau = \tan\theta$. It is clear that μ_c and τ meet all requirements of Theorem 1.3.

Next we fix the lattice Γ such that the dual lattice is spanned by the wave vectors $(1, \pm\tau)$, and set

$$E_s = F_s = H^s_{o,e}(\mathbb{R}^2/\Gamma) \times H^s_{e,e}(\mathbb{R}^2/\Gamma).$$

Since the scaling mapping $u \to u \circ \mathbb{T}^{-1}$ establishes an isomorphism between $H^s(\mathbb{R}^2/\Gamma)$ and a closed subspace of the Sobolev space H^s of doubly 2π–periodic functions, the properties $(A1)$ and $(A2)$ are clear. A smoothing operator with the required properties can be defined by

$$S_\wp u = \frac{\sqrt{\tau}}{2\pi} \sum_{K \in \Gamma'} \varsigma(\wp|K|)\hat{u}_K e^{iK \cdot X},$$

where $\varsigma : \mathbb{R}^+ \mapsto \mathbb{R}^+$ is a smooth function which equals 1 on $[0,1]$ and 0 on $[2, \infty)$.

Fix an arbitrary $N > 3$ and define the operator $\Phi = (\Phi_1, \Phi_2)$ by the equalities

(6.12) $\qquad \Phi(W, \varepsilon) = \varepsilon^{-N}\mathcal{F}\bigl(U_\varepsilon^{(2N)} + \varepsilon^N W, \mu_\varepsilon^{(2N)}\bigr),$

where $(U_\varepsilon^{(2N)}, \mu_\varepsilon^{(2N)})$ is the approximate solution at order ε^{2N} obtained at Theorem 2.3 in the case of diamond waves ($\varepsilon_1 = \varepsilon_2 = \varepsilon/2$), and $\tau = \tan\theta$, $\mu_c = \cos\theta$, $\delta < \tau < 1/\delta$. By construction $W_\varepsilon^{(N)} = 0$, hence (6.10) and (6.11) are satisfied.

It follows from Lemma 1.1 that the continuous mapping $(W, \varepsilon) \mapsto \Phi(W, \varepsilon)$ from $E_{s+1} \times \mathbb{R}$ into F_s for $s \geq 2$, is of class of C^∞. Applying the same arguments as in section 9 of [**IPT**] we conclude from this that for $5 \leq \rho \leq r - 18$, the operator Φ satisfies Conditions (B1) and (B2), and inequality (6.7) from Condition (B4). Let us denote by Λ the approximate differential, defined as

$$\Lambda(W, \varepsilon) = \mathcal{L}(U_\varepsilon^{(2N)} + \varepsilon^N W, \mu_\varepsilon^{(2N)}) \begin{pmatrix} 1 & -\mathfrak{b} \\ 0 & 1 \end{pmatrix}$$

where the operator $\mathcal{L}$ and coefficient $\mathfrak{b}$ are defined by formulae (3.1) and (3.3). It follows from this and (3.2) that the linear operator

$$\mathcal{R} = \Lambda(W, \varepsilon) - \partial_W \Phi(W, \varepsilon),$$

is defined by $\mathcal{R}\delta W = (R_1 \delta W, 0)$ with

$$R_1 \delta W = \mathcal{G}_\eta \left(\frac{\Phi_1 \delta \eta}{1 + (\nabla \eta)^2} \right) - \nabla \cdot \left(\frac{\Phi_1 \delta \eta}{1 + (\nabla \eta)^2} \nabla \eta \right),$$

where $\delta W = (\delta\psi, \delta\eta)$ and $\Phi_1 = \Phi_1(W, \varepsilon)$. Then, thanks to Lemma 3.1, we have for $s \geq 2$, and $\|U\|_3 \leq M_3$

$$\|\Lambda(W, \varepsilon)u\|_s \leq c(M_3)(1 + \|W\|_{s+2})\|u\|_{s+1},$$

$$\|\mathcal{R}u\|_s \leq c_s(M_3)\{\|\Phi\|_2(1 + \|W\|_{s+2})\|u\|_2 + \|u\|_{s+1}) + \|\Phi\|_{s+1}\|u\|_2.$$

Therefore, the operators Φ and Λ satisfy Conditions (B3) and (B4). Now taking into account the result of Theorem 5.9, it appears that Condition (B5) is satisfied for

$$5 \leq \rho \leq r - 18$$
$$\sigma = 3, \quad \varrho = 2,$$

Finally, since $N \geq 3$, condition (B6) is also satisfied thanks to Theorem 4.2, so we can apply Theorem 6.1, and Theorem 1.3 is proved.

APPENDIX A

Analytical study of $\mathcal{G}_\eta$

A.1. Computation of the differential of $\mathcal{G}_\eta$

Let us make the following change of coordinate for x_3
$$s = x_3 - \eta(X)$$
and denote by θ the function defined by
$$\theta(X, s) = \varphi(X, s + \eta(X)).$$
Then the Dirichlet-Neumann operator is defined by

(A.1) $\quad \Delta\theta - 2\nabla_X\eta \cdot \nabla_X(\dfrac{\partial\theta}{\partial s}) - \dfrac{\partial\theta}{\partial s}\Delta_X\eta + \dfrac{\partial^2\theta}{\partial s^2}(\nabla_X\eta)^2 = 0, \quad s < 0$

$$\theta|_{s=0} = \psi$$
$$\nabla\theta \to 0 \text{ as } s \to -\infty,$$

and

(A.2) $\quad \mathcal{G}_\eta\psi = (1 + (\nabla_X\eta)^2)\dfrac{\partial\theta}{\partial s}|_{s=0} - \nabla_X\eta \cdot \nabla_X\psi.$

Notice that the above equations (A.1), (A.2) may be written into the form
$$\nabla \cdot (P_\eta\nabla\theta) = 0, \quad s < 0,$$
$$\mathcal{G}_\eta\psi = (P_\eta\nabla\theta) \cdot \mathbf{e}_3$$
where $\mathbf{e}_3$ is the unit vertical vector, and P_η is the following symmetric matrix
$$P_\eta = \begin{pmatrix} \mathbb{I} & -\nabla_X\eta \\ -(\nabla_X\eta)^t & 1 + (\nabla_X\eta)^2 \end{pmatrix}.$$

We can see easily that the operator $\mathcal{G}_\eta$ is *symmetric and non negative in* $L^2(\mathbb{R}^2/\Gamma)$: for η, ψ_1, ψ_2 smooth enough bi-periodic functions

$$\langle \mathcal{G}_\eta\psi_1, \psi_2 \rangle = \langle (P_\eta\nabla\theta_1) \cdot \mathbf{e}_3, \psi_2 \rangle$$
$$= \int_{-\infty}^{0}\int_{\mathbb{R}^2/\Gamma} (\nabla\theta_2 \cdot P_\eta\nabla\theta_1)dXds$$

which is symmetric. Moreover, we have
$$\nabla\theta \cdot P_\eta\nabla\theta = (\nabla_X\theta - \nabla_X\eta\dfrac{\partial\theta}{\partial s})^2 + (\dfrac{\partial\theta}{\partial s})^2 \geq 0,$$
hence, for η and ψ smooth enough bi-periodic functions
$$\langle \mathcal{G}_\eta\psi, \psi \rangle \geq 0.$$

Let us now compute formally the differential of $\mathcal{G}_\eta$. From (A.2), we obtain

(A.3) $\qquad \partial_\eta \mathcal{G}_\eta[h]\psi = (P_\eta \nabla \theta_1) \cdot \mathbf{e}_3 + (Q_\eta[h]\nabla \theta) \cdot \mathbf{e}_3$

with θ as above, and where θ_1 satisfies the system

$$\begin{aligned}
\nabla \cdot (P_\eta \nabla \theta_1) &= -\nabla \cdot (Q_\eta[h]\nabla \theta), \quad s < 0 \\
\theta_1|_{s=0} &= 0, \\
\nabla \theta_1 &\to 0, \quad s \to -\infty,
\end{aligned}$$

$$Q_\eta[h] = \begin{pmatrix} 0 & -\nabla_X h \\ -(\nabla_X h)^t & 2\nabla_X \eta \cdot \nabla_X h \end{pmatrix}.$$

Let us notice from (A.1), (A.2), that we have

$$\nabla \cdot \left(P_\eta \nabla (h\frac{\partial \theta}{\partial s})\right) + \nabla \cdot (Q_\eta[h]\nabla \theta) = 0, \quad s < 0.$$

Indeed, we have

$$\begin{aligned}
P_\eta \nabla (h\frac{\partial \theta}{\partial s}) &= h P_\eta (\nabla \frac{\partial \theta}{\partial s}) + \frac{\partial \theta}{\partial s}\begin{pmatrix} \nabla_X h \\ -\nabla_X \eta \cdot \nabla_X h \end{pmatrix}, \\
Q_\eta[h]\nabla \theta &= \begin{pmatrix} -\nabla_X h \frac{\partial \theta}{\partial s} \\ -\nabla_X \theta \cdot \nabla_X h + 2(\nabla_X \eta \cdot \nabla_X h)\frac{\partial \theta}{\partial s} \end{pmatrix},
\end{aligned}$$

hence

$$\nabla \cdot \left(P_\eta \nabla (h\frac{\partial \theta}{\partial s})\right) + \nabla \cdot (Q_\eta[h]\nabla \theta) =$$

$$= h\nabla \cdot (P_\eta \nabla \frac{\partial \theta}{\partial s}) + \nabla_X h \cdot \left\{(P_\eta \nabla \frac{\partial \theta}{\partial s}) - \nabla_X \frac{\partial \theta}{\partial s} + \nabla_X \eta \frac{\partial^2 \theta}{\partial s^2}\right\},$$

and this cancels, thanks to $\nabla \cdot (P_\eta \nabla \theta) = 0$, and to the definition of P_η.

It results that

$$\begin{aligned}
\theta_1 &= h\frac{\partial \theta}{\partial s} + \theta_2, \\
\nabla \cdot (P_\eta \nabla \theta_2) &= 0, \quad s < 0, \\
\theta_2|_{s=0} &= -h\frac{\partial \theta}{\partial s}|_{s=0}, \\
\nabla \theta_2 &\to 0, \quad s \to -\infty,
\end{aligned}$$

hence looking at relationship (A.3), we obtain

$$\begin{aligned}
\partial_\eta \mathcal{G}_\eta[h]\psi &= (P_\eta \nabla \theta_2) \cdot \mathbf{e}_3 + (P_\eta \nabla (h\frac{\partial \theta}{\partial s})) \cdot \mathbf{e}_3 + (Q_\eta[h]\nabla \theta) \cdot \mathbf{e}_3 \\
&= -\mathcal{G}_\eta(h\frac{\partial \theta}{\partial s}|_0) + h\{(1 + (\nabla_X \eta)^2)\frac{\partial^2 \theta}{\partial s^2}|_0 - \nabla_X \eta \cdot \nabla_X \frac{\partial \theta}{\partial s}|_0\} + \\
&\quad + \nabla_X h \cdot \{\nabla_X \eta \frac{\partial \theta}{\partial s}|_0 - \nabla_X \psi\},
\end{aligned}$$

and using $\nabla \cdot (P_\eta \nabla \theta) = 0$, we get

$$\partial_\eta \mathcal{G}_\eta[h]\psi = -\mathcal{G}_\eta(h\frac{\partial \theta}{\partial s}|_0) + h\{-\Delta_X\psi + \nabla_X\eta \cdot \nabla_X\frac{\partial \theta}{\partial s}|_0 + \Delta_X\eta\frac{\partial \theta}{\partial s}|_0\} +$$
$$+\nabla_X h \cdot \{\nabla_X\eta\frac{\partial \theta}{\partial s}|_0 - \nabla_X\psi\}$$
$$= -\mathcal{G}_\eta(h\frac{\partial \theta}{\partial s}|_0) + \nabla_X \cdot \{h(\nabla_X\eta\frac{\partial \theta}{\partial s}|_0 - \nabla_X\psi)\}$$

as required in Lemma 2.1, thanks to (A.2).

A.2. Second order Taylor expansion of $\mathcal{G}_\eta$ in $\eta = 0$

Let us consider the Taylor expansion of $\mathcal{G}_\eta$ in "powers" of η:

$$\mathcal{G}_\eta \psi = \mathcal{G}^{(0)}\psi + \mathcal{G}^{(1)}\{\eta\}\psi + \mathcal{G}^{(2)}\{\eta^{(2)}\}\psi + ...$$

where $\mathcal{G}^{(k)}$ is $k-$ linear symmetric with respect to η, and linear in ψ. Moreover, for any $k \geq 1$ and $m \geq 2$, $\mathcal{G}^{(k)}$ is bounded from

$$\{H^{m+1}(\mathbb{R}^2/\Gamma)\}^k \quad \text{into} \quad \mathcal{L}(H_0^{m+1}(\mathbb{R}^2/\Gamma), H_0^m(\mathbb{R}^2/\Gamma)).$$

From the lemma above and (2.1), (2.2), we obtain

(A.4) $$\mathcal{G}^{(1)}\{\eta\}\psi = -\mathcal{G}^{(0)}(\eta \mathcal{G}^{(0)}\psi) - \nabla \cdot (\eta \nabla \psi).$$

Differentiating once more (2.2) with respect to η, we now obtain in 0

$$\partial_\eta \zeta[h] = -\mathcal{G}^{(0)}(h\mathcal{G}^{(0)}\psi) - h\Delta\psi,$$

and differentiating (2.1) with respect to η, we then obtain the (symmetric) second order derivative in 0, and

(A.5) $$\mathcal{G}^{(2)}\{\eta^{(2)}\}\psi = \mathcal{G}^{(0)}(\eta \mathcal{G}^{(0)}(\eta \mathcal{G}^{(0)}\psi)) + \frac{1}{2}\mathcal{G}^{(0)}(\eta^2 \Delta\psi) + \frac{1}{2}\Delta(\eta^2 \mathcal{G}^{(0)}\psi).$$

Now, if we Fourier expand any bi-periodic function as

$$\psi = \sum_{K \in \Gamma'} \psi_K e^{iK \cdot X}$$

then we have

(A.6) $$\{\mathcal{G}^{(0)}\psi\}_K = |K|\psi_K,$$

(A.7) $$\{\mathcal{G}^{(1)}\{\eta\}\psi\}_K = \sum_{K_1+K_2=K,\ K_j \in \Gamma'} \{(K \cdot K_1) - |K||K_1|\}\psi_{K_1}\eta_{K_2},$$

(A.8) $$\{\mathcal{G}^{(2)}\{\eta^{(2)}\}\psi\}_K = \sum_{K_1+K_2+K_3=K,\ K_j \in \Gamma'} \frac{|K||K_1|}{2}\{|K_1+K_2| + |K_1+K_3| - |K| - |K_1|\}\psi_{K_1}\eta_{K_2}\eta_{K_3},$$

and we check that, for $m \geq 2$, the operators are bounded from $\{H^{m+1}(\mathbb{R}^2/\Gamma)\}^k$ into $\mathcal{L}(H_0^{m+1}(\mathbb{R}^2/\Gamma), H_0^m(\mathbb{R}^2/\Gamma))$ since we have

$$||(K_1+K_2) \cdot K_1| - |K_1+K_2||K_1|| \leq 4|K_1||K_2|,$$

and there exists a constant c such that

$$|K_1 + K_2 + K_3|||K_1 + K_2| + |K_1 + K_3| - |K_1 + K_2 + K_3| - |K_1|| \le$$
$$\le c|K_2||K_3|.$$

APPENDIX B

Formal computation of 3-dimensional waves

In taking $\mathbf{u}_0 = (1,0)$, the symmetric linearized operator for $\mu = \mu_c$ and $\mathbf{u} = \mathbf{u}_0$ reads

$$\mathcal{L}_0 = \begin{pmatrix} \mathcal{G}^{(0)} & -\frac{\partial}{\partial x_1} \\ \frac{\partial}{\partial x_1} & \mu_c \end{pmatrix},$$

and $\mathcal{L}_0$ has a four-dimensional kernel, spanned by the vectors

$$\begin{aligned}
\zeta_0 &= (1, -i/\mu_c)e^{iK_1 \cdot X}, \quad \overline{\zeta}_0 = (1, i/\mu_c)e^{-iK_1 \cdot X}, \\
\zeta_1 &= (1, -i/\mu_c)e^{iK_2 \cdot X}, \quad \overline{\zeta}_1 = (1, i/\mu_c)e^{-iK_2 \cdot X}.
\end{aligned}$$

We observe that the action of different symmetries of the system on eigenvectors is as follows:

$$\begin{aligned}
\mathcal{T}_\mathbf{v}\zeta_0 &= \zeta_0 e^{iK_1 \cdot \mathbf{v}}, \quad \mathcal{T}_\mathbf{v}\zeta_1 = \zeta_1 e^{iK_2 \cdot \mathbf{v}}, \\
\mathcal{S}_0\zeta_0 &= -\overline{\zeta}_0, \quad \mathcal{S}_0\zeta_1 = -\overline{\zeta}_1, \\
\mathcal{S}_1\zeta_0 &= \zeta_1, \quad \mathcal{S}_1\zeta_1 = \zeta_0.
\end{aligned}$$

Let us write formally the nonlinear system (1.6), (1.7) under the form
(B.1)
$$\mathcal{L}_0 U + \tilde{\mu}\mathcal{L}_1 U + \mathcal{L}_2(\omega, U) + \mathcal{N}_2(U,U) + \mathcal{N}_3(U,U,U) + O(\|U\|^4 + |\omega|\|U\|^2) = 0,$$

with $U = (\psi, \eta)$, $\tilde{\mu} = \mu - \mu_c$, $\omega = \mathbf{u} - \mathbf{u}_0$

$$\begin{aligned}
\mathcal{L}_1 U &= (0, \eta), \\
\mathcal{L}_2(\omega, U) &= (-\omega \cdot \nabla \eta, \omega \cdot \nabla \psi),
\end{aligned}$$

$$\mathcal{N}_2(U,U) = \left\{ \begin{array}{l} \mathcal{G}^{(1)}\{\eta\}\psi \\ \frac{1}{2}\nabla\psi^2 - \frac{1}{2}(\frac{\partial\eta}{\partial x_1})^2 \end{array} \right. ,$$

$$\mathcal{N}_3(U,U,U) = \left\{ \begin{array}{l} \mathcal{G}^{(2)}\{\eta,\eta\}\psi \\ -\frac{\partial\eta}{\partial x_1}(\nabla\eta \cdot \nabla\psi) \end{array} \right. .$$

B.1. Formal Fredholm alternative

Let us consider the formal resolution of the linear system

$$\mathcal{L}_0 U = F = (f, g),$$

with

$$U = \sum_{n=(n_1,n_2)\in\mathbb{Z}^2} U_n e^{i(n_1 K_1 \cdot X + n_2 K_2 \cdot X)}, \quad U_n = (\psi_n, \eta_n), \quad \psi_0 = 0,$$

$$F = \sum_{n=(n_1,n_2)\in\mathbb{Z}^2} F_n e^{i(n_1 K_1 \cdot X + n_2 K_2 \cdot X)}, \quad F_n = (f_n, g_n), \quad f_0 = 0.$$

Then, we have

$$|n_1 K_1 + n_2 K_2|\psi_n - i\{(n_1 K_1 + n_2 K_2) \cdot \mathbf{u}_0\}\eta_n = f_n$$
$$i\{(n_1 K_1 + n_2 K_2) \cdot \mathbf{u}_0\}\psi_n + \mu_c \eta_n = g_n,$$

hence for

$$\{(n_1 K_1 + n_2 K_2) \cdot \mathbf{u}_0\}^2 - \mu_c |n_1 K_1 + n_2 K_2| \neq 0$$

i.e. by assumption for $(n_1, n_2) \neq (\pm 1, 0), (0, \pm 1)$, this leads to

$$(B.2) \quad \psi_n = -\frac{\mu_c f_n + i\{(n_1 K_1 + n_2 K_2) \cdot \mathbf{u}_0\}g_n}{\{(n_1 K_1 + n_2 K_2) \cdot \mathbf{u}_0\}^2 - \mu_c |n_1 K_1 + n_2 K_2|},$$

$$(B.3) \quad \eta_n = \frac{i\{(n_1 K_1 + n_2 K_2) \cdot \mathbf{u}_0\}f_n - |n_1 K_1 + n_2 K_2|g_n}{\{(n_1 K_1 + n_2 K_2) \cdot \mathbf{u}_0\}^2 - \mu_c |n_1 K_1 + n_2 K_2|},$$

and for $(n_1, n_2) = (0, 0)$

$$\psi_{0,0} = 0, \quad \eta_{0,0} = \frac{1}{\mu_c} g_{0,0},$$

while for $(n_1, n_2) = (\pm 1, 0), (0, \pm 1)$, we need to satisfy the compatibility conditions

$$(F, \zeta_0) = (F, \overline{\zeta}_0) = (F, \zeta_1) = (F, \overline{\zeta}_1) = 0$$

which gives

$$\mu_c f_{1,0} + i g_{1,0} = 0,$$
$$\mu_c f_{-1,0} - i g_{-1,0} = 0,$$
$$\mu_c f_{0,1} + i g_{0,1} = 0,$$
$$\mu_c f_{0,-1} - i g_{0,-1} = 0.$$

For uniqueness of the definition of the pseudo-inverse $\widetilde{\mathcal{L}}_0^{-1}$, we fix U such that

$$(U, \zeta_0) = (U, \overline{\zeta}_0) = (U, \zeta_1) = (U, \overline{\zeta}_1) = 0,$$

hence this leads to

$$\psi_{1,0} = -i\frac{1+\tau^2}{2+\tau^2}g_{1,0} = \frac{1}{\mu_c(2+\tau^2)}f_{1,0},$$

$$\eta_{1,0} = \frac{1}{\mu_c(2+\tau^2)}g_{1,0} = \frac{i}{2+\tau^2}f_{1,0},$$

$$\psi_{0,1} = -i\frac{1+\tau^2}{2+\tau^2}g_{0,1} = \frac{1}{\mu_c(2+\tau^2)}f_{0,1},$$
$$\eta_{0,1} = \frac{1}{\mu_c(2+\tau^2)}g_{0,1} = \frac{i}{2+\tau^2}f_{0,1}.$$

B.2. Bifurcation equation

Now coming back to (B.1), we use a *formal* Lyapunov - Schmidt method and decompose U as follows

$$U = W + V$$
$$W = A\zeta_0 + \overline{A}\overline{\zeta}_0 + B\zeta_1 + \overline{B}\overline{\zeta}_1 = \mathcal{P}_0 U$$
$$(V, \zeta_0) = (V, \overline{\zeta}_0) = (V, \zeta_1) = (V, \overline{\zeta}_1) = 0.$$

We can solve formally with respect to V the part of equ. (B.1) orthogonal to the 4-dimensional kernel of $\mathcal{L}_0$, as a uniquely determined formal power series in $\omega, \tilde{\mu}, A, \overline{A}, B, \overline{B}$, which we write as

$$V = \mathcal{V}(\tilde{\mu}, \omega, A, \overline{A}, B, \overline{B}).$$

The uniqueness of the series and the symmetries of our system lead to the following identities

$$\mathcal{T}_{\mathbf{v}}\mathcal{V}(\tilde{\mu}, \omega, A, \overline{A}, B, \overline{B}) = \mathcal{V}(\tilde{\mu}, \omega, Ae^{iK_1 \cdot \mathbf{v}}, \overline{A}e^{-iK_1 \cdot \mathbf{v}}, Be^{iK_2 \cdot \mathbf{v}}, \overline{B}e^{-iK_2 \cdot \mathbf{v}}),$$
$$\mathcal{S}_0 \mathcal{V}(\tilde{\mu}, \omega, A, \overline{A}, B, \overline{B}) = \mathcal{V}(\tilde{\mu}, \omega, -\overline{A}, -A, -\overline{B}, -B)$$
$$\mathcal{S}\mathcal{V}(\tilde{\mu}, \mathbf{0}, A, \overline{A}, B, \overline{B}) = \mathcal{V}(\tilde{\mu}, \mathbf{0}, B, \overline{B}, A, \overline{A}).$$

The principal part of V is given by

$$V = -\widetilde{\mathcal{L}}_0^{-1}(\mathbb{I} - \mathcal{P}_0)\mathcal{N}_2(W, W) + O\{(|\tilde{\mu}| + |\omega|)\|W\| + \|W\|^3\},$$
$$-\widetilde{\mathcal{L}}_0^{-1}\mathcal{N}_2(W, W) = A^2 U_{2000} + |A|^2 U_{1100} + \overline{A}^2 U_{0200} + AB U_{1010} + \overline{A}B U_{0110} +$$
$$+ A\overline{B} U_{1001} + \overline{A}\overline{B} U_{0101} + B^2 U_{0020} + |B|^2 U_{0011} + \overline{B}^2 U_{0002}$$

where we observe easily that

$$\mathcal{P}_0 \mathcal{N}_2(W, W) = 0.$$

Using (A.7), and

(B.4) $$2\mathcal{N}_2(U_1, U_2) = \begin{cases} \mathcal{G}^{(1)}\{\eta_1\}\psi_2 + \mathcal{G}^{(1)}\{\eta_2\}\psi_1 \\ \nabla\psi_1 \cdot \nabla\psi_2 - \frac{\partial\eta_1}{\partial x_1}\frac{\partial\eta_2}{\partial x_1} \end{cases},$$

we find

$$\mathcal{N}_2(W, W) = A^2 V_{2000} + |A|^2 V_{1100} + \overline{A}^2 V_{0200} + AB V_{1010} + \overline{A}B V_{0110} +$$
$$+ A\overline{B} V_{1001} + \overline{A}\overline{B} V_{0101} + B^2 V_{0020} + |B|^2 V_{0011} + \overline{B}^2 V_{0002}$$

with

$$V_{2000} = (0, -\frac{1}{\mu_c^2})e^{2iK_1 \cdot X}, \quad V_{0020} = (0, -\frac{1}{\mu_c^2})e^{2iK_2 \cdot X}$$

$$V_{1100} = 0, \quad V_{0011} = 0, \quad V_{0200} = \overline{V}_{2000}, \quad V_{0002} = \overline{V}_{0020}$$

$$V_{1010} = (\frac{4i}{\mu_c^2}(1-\mu_c), -2)e^{i(K_1 \cdot X + K_2 \cdot X)}, \quad V_{0101} = \overline{V}_{1010}$$

$$V_{0110} = (0, -2\tau^2)e^{i(-K_1 \cdot X + K_2 \cdot X)}, \quad V_{1001} = \overline{V}_{0110}.$$

Now, thanks to (B.2), (B.3) we obtain $-\widetilde{\mathcal{L}}_0^{-1}\mathcal{N}_2(W,W)$ as follows

$$U_{2000} = \left(\frac{-i}{\mu_c^2}, -\frac{1}{\mu_c^3}\right)e^{2iK_1 \cdot X}, \quad U_{0020} = \left(\frac{-i}{\mu_c^2}, -\frac{1}{\mu_c^3}\right)e^{2iK_2 \cdot X},$$

$$U_{1100} = 0, \quad U_{0011} = 0, \quad U_{0200} = \overline{U}_{2000}, \quad U_{0002} = \overline{U}_{0020},$$

$$U_{1010} = \left(\frac{-2i(2\mu_c - 1)}{\mu_c(2-\mu_c)}, \frac{-2(\mu_c^2 + 2\mu_c - 2)}{\mu_c^2(2-\mu_c)}\right)e^{i(K_1 \cdot X + K_2 \cdot X)}, \quad U_{0101} = \overline{U}_{1010},$$

$$U_{0110} = \left(0, \frac{2\tau^2}{\mu_c}\right)e^{i(-K_1 \cdot X + K_2 \cdot X)}, \quad U_{1001} = \overline{U}_{0110}.$$

Replacing V by $\mathcal{V}(\tilde{\mu}, \omega, A, \overline{A}, B, \overline{B})$ in the compatibility conditions, i.e. the components of (B.1) on ker $\mathcal{L}_0$,

$$\langle \tilde{\mu}\mathcal{L}_1(W+\mathcal{V}) + \mathcal{L}_2(\omega, W+\mathcal{V}) + \mathcal{N}_2(W+\mathcal{V}, W+\mathcal{V}) + ..., \zeta_0 \rangle = 0,$$
$$\langle \tilde{\mu}\mathcal{L}_1(W+\mathcal{V}) + \mathcal{L}_2(\omega, W+\mathcal{V}) + \mathcal{N}_2(W+\mathcal{V}, W+\mathcal{V}) + ..., \zeta_1 \rangle = 0,$$

noticing that the complex conjugate equations are then automatically satisfied, lead to two complex equations of the form

$$f(\tilde{\mu}, \omega, A, \overline{A}, B, \overline{B}) = 0,$$
$$g(\tilde{\mu}, \omega, A, \overline{A}, B, \overline{B}) = 0,$$

for which the equivariance of the system (B.1) with respect to various symmetries leads to the following properties

$$f(\tilde{\mu}, \omega, Ae^{iK_1 \cdot \mathbf{v}}, \overline{A}e^{-iK_1 \cdot \mathbf{v}}, Be^{iK_2 \cdot \mathbf{v}}, \overline{B}e^{-iK_2 \cdot \mathbf{v}}) = e^{iK_1 \cdot \mathbf{v}}f(\tilde{\mu}, \omega, A, \overline{A}, B, \overline{B})$$
$$g(\tilde{\mu}, \omega, Ae^{iK_1 \cdot \mathbf{v}}, \overline{A}e^{-iK_1 \cdot \mathbf{v}}, Be^{iK_2 \cdot \mathbf{v}}, \overline{B}e^{-iK_2 \cdot \mathbf{v}}) = e^{iK_2 \cdot \mathbf{v}}g(\tilde{\mu}, \omega, A, \overline{A}, B, \overline{B})$$
$$f(\tilde{\mu}, \omega, -\overline{A}, -A, -\overline{B}, -B) = -\overline{f}(\tilde{\mu}, \omega, A, \overline{A}, B, \overline{B})$$
$$g(\tilde{\mu}, \omega, -\overline{A}, -A, -\overline{B}, -B) = -\overline{g}(\tilde{\mu}, \omega, A, \overline{A}, B, \overline{B})$$
$$f(\tilde{\mu}, \mathbf{0}, B, \overline{B}, A, \overline{A}) = g(\tilde{\mu}, \mathbf{0}, A, \overline{A}, B, \overline{B}).$$

Since K_1 and K_2 are linearly independent, it results that f and g take formally the form

$$f(\tilde{\mu}, \omega, A, \overline{A}, B, \overline{B}) = A\phi_0(\tilde{\mu}, \omega, |A|^2, |B|^2)$$
$$g(\tilde{\mu}, \omega, A, \overline{A}, B, \overline{B}) = B\phi_1(\tilde{\mu}, \omega, |A|^2, |B|^2),$$

where functions ϕ_0 and ϕ_1 are real valued. Moreover, for $\omega = 0$ we have

$$\phi_1(\tilde{\mu}, \mathbf{0}, |A|^2, |B|^2) = \phi_0(\tilde{\mu}, \mathbf{0}, |B|^2, |A|^2).$$

It results immediately that we have the following (formal) solutions of our system (in addition to the trivial solution 0):

PROOF. i) $B = 0$, $|A|$ satisfying $\phi_0(\tilde{\mu}, \omega, |A|^2, 0) = 0$, which is not else that the 2-dimensional travelling wave with basic wave vector K_1, and where, with no loss of generality, we can choose the velocity $\mathbf{c}$ in the direction of K_1.

ii) $A = 0$, $|B|$ satisfying $\phi_1(\tilde{\mu}, \omega, 0, |B|^2) = 0$, which is not else that the 2-dimensional travelling wave with basic wave vector K_2, and where, with no loss of generality, we can choose the velocity $\mathbf{c}$ in the direction of K_2.

iii) $|A|$ and $|B|$ such that

(B.5)
$$\phi_0(\tilde{\mu}, \omega, |A|^2, |B|^2) = 0,$$
$$\phi_1(\tilde{\mu}, \omega, |A|^2, |B|^2) = 0,$$

which gives a family of 3-dimensional travelling waves. Moreover we notice that if $\omega = 0$ there is a family of solutions where $|A| = |B|$ and

$$\phi_0(\tilde{\mu}, \mathbf{0}, |A|^2, |A|^2) = 0,$$

representing the "diamond waves" of [**ReSh**]. The leading terms of ϕ_0 and ϕ_1 are computed by Bridges, Dias, Menasce in [**BDM**], even in cases with a finite depth and with surface tension. Since our case has less parameters, our computations may look simpler. The leading terms independent of $|A|$ and $|B|$ in ϕ_0 come from

$$\langle \tilde{\mu} \mathcal{L}_1 \zeta_0 + \mathcal{L}_2(\omega, \zeta_0), \zeta_0 \rangle = \frac{4\pi^2}{\tau} \frac{1}{\mu_c} \left(\frac{\tilde{\mu}}{\mu_c} - 2\omega \cdot K_1 \right),$$

and we notice that (in using $\frac{\delta \mu}{\mu} = -2\frac{\delta c}{c}$)

$$\frac{\tilde{\mu}}{\mu_c} - 2\omega \cdot K_1 \sim -2\frac{(\mathbf{c} - \mathbf{c}_0) \cdot K_1}{c_0}.$$

We then have

$$\phi_0(\tilde{\mu}, \omega, |A|^2, |B|^2) = \frac{1}{\mu_c} \left(\frac{\tilde{\mu}}{\mu_c} - 2\omega \cdot K_1 \right) +$$
$$\alpha_0 |A|^2 + \beta_0 |B|^2 + O\{(|\tilde{\mu}| + |\omega| + |A|^2 + |B|^2)^2\}$$
$$\phi_1(\tilde{\mu}, \omega, |A|^2, |B|^2) = \frac{1}{\mu_c} \left(\frac{\tilde{\mu}}{\mu_c} - 2\omega \cdot K_2 \right) +$$
$$\beta_0 |A|^2 + \alpha_0 |B|^2 + O\{(|\tilde{\mu}| + |\omega| + |A|^2 + |B|^2)^2\},$$

with

$$\alpha_0 = \frac{\tau}{4\pi^2} \langle 2\mathcal{N}_2(\overline{\zeta}_0, U_{2000}) + 3\mathcal{N}_3(\zeta_0, \zeta_0, \overline{\zeta}_0), \zeta_0 \rangle,$$
$$\beta_0 = \frac{\tau}{4\pi^2} \langle 2\mathcal{N}_2(\zeta_1, U_{1001}) + 2\mathcal{N}_2(\overline{\zeta}_1, U_{1010}) + 6\mathcal{N}_3(\zeta_0, \zeta_1, \overline{\zeta}_1), \zeta_0 \rangle.$$

We can formally solve the system of equations
$$\phi_0(\tilde{\mu}, \omega, |A|^2, |B|^2) = 0,$$
$$\phi_1(\tilde{\mu}, \omega, |A|^2, |B|^2) = 0,$$
with respect to $\tilde{\mu}, \omega_2 = \frac{1}{2\tau}\omega \cdot (K_1 - K_2)$. Indeed, we obtain respectively in adding and subtracting the two equations, a system easy to solve, in taking into account of $\omega_1 = O(\omega_2^2)$,

$$\tilde{\mu} = -\frac{\mu_c^2}{2}(\alpha_0 + \beta_0)(|A|^2 + |B|^2) + O\{(|A|^2 + |B|^2)^2\},$$

$$\omega_2 = (|A|^2 - |B|^2)\left(\frac{\mu_c}{4\tau}(\alpha_0 - \beta_0) + O\{|A|^2 + |B|^2\}\right).$$

For the computation of coefficients α_0 and β_0 we use again (A.7) and (B.4) and obtain

$$2\mathcal{N}_2(\overline{\zeta}_0, U_{2000}) = \left(\frac{2}{\mu_c^5}, 0\right) e^{iK_1 \cdot X},$$

$$2\mathcal{N}_2(\zeta_1, U_{1001}) = \left(-\frac{4\tau^4}{\mu_c}, 0\right) e^{iK_1 \cdot X},$$

$$2\mathcal{N}_2(\overline{\zeta}_1, U_{1010}) = \left(\frac{4(\mu_c^3 + 4\mu_c^2 - 5\mu_c + 1)}{\mu_c^3(2 - \mu_c)}, \frac{8i(\mu_c - 1)(1 - \mu_c^2)}{\mu_c^3(2 - \mu_c)}\right) e^{iK_1 \cdot X}.$$

Now, with (A.8) and the form of $\mathcal{N}_3$ we have

$$3\mathcal{N}_3(\zeta_0, \zeta_0, \overline{\zeta}_0) = \left(\frac{1}{\mu_c^5}, \frac{-i}{\mu_c^4}\right) e^{iK_1 \cdot X},$$

$$6\mathcal{N}_3(\zeta_0, \zeta_1, \overline{\zeta}_1) = \left(\frac{2}{\mu_c^4}(2 - \frac{1}{\mu_c}), \frac{-2i}{\mu_c^2}(2 - \frac{1}{\mu_c^2})\right) e^{iK_1 \cdot X}.$$

Finally we obtain

$$\alpha_0 = \frac{4}{\mu_c^5}$$

$$\beta_0 = -\frac{8}{\mu_c^5} + \frac{8}{\mu_c^4} + \frac{12}{\mu_c^3} - \frac{32}{\mu_c^2} - \frac{8}{\mu_c} + \frac{36}{\mu_c^2(2 - \mu_c)}.$$

If $\alpha_0 + \beta_0$ and $\alpha_0 - \beta_0$ are both different from 0, we can solve the system (B.5) with respect to $|A|^2$ and $|B|^2$ and obtain a formal expansion in powers of $(\tilde{\mu}, \omega)$ of the form

$$|A|^2 = \frac{\tilde{\mu}}{-\mu_c^2(\alpha_0 + \beta_0)} + \frac{\omega \cdot (K_1 - K_2)}{\mu_c(\alpha_0 - \beta_0)} + O(|(\tilde{\mu}, \omega)|^2),$$

$$|B|^2 = \frac{\tilde{\mu}}{-\mu_c^2(\alpha_0 + \beta_0)} - \frac{\omega \cdot (K_1 - K_2)}{\mu_c(\alpha_0 - \beta_0)} + O(|(\tilde{\mu}, \omega)|^2).$$

The indeterminacy on the phases of A and B means that we obtain in fact, for each fixed $(\tilde{\mu}, \omega)$ leading to positive expressions for $|A|^2$ and $|B|^2$, a torus of solutions, which is generated by acting the operator $\mathcal{T}_\mathbf{v}$ on a particular solution, for instance with A and B pure imaginary. This two-parameter family of tori of 3-dimensional waves connects with the 2-dimensional travelling waves respectively of wave vectors K_1 and K_2 (choosing ω orthogonal to K_2 or to K_1. If we choose $\omega = 0$ which means that we choose the direction of the waves as x_1 axis, then we obtain the "diamond waves" as in [**ReSh**], and [**BDM**], here without surface tension.

Let us study the sign of $(\alpha_0 + \beta_0)$ and $(\alpha_0 - \beta_0)$. We have

$$\alpha_0 + \beta_0 = \frac{4}{\mu_c^2}\left(-\frac{1}{\mu_c^3} + \frac{2}{\mu_c^2} + \frac{3}{\mu_c} - 8 - 2\mu_c + \frac{9}{2-\mu_c}\right)$$

$$\beta_0 - \alpha_0 = \frac{4}{\mu_c^2}\left(-\frac{3}{\mu_c^3} + \frac{2}{\mu_c^2} + \frac{3}{\mu_c} - 8 - 2\mu_c + \frac{9}{2-\mu_c}\right),$$

where we notice that

$$\alpha_0 + \beta_0 = \frac{4}{\mu_c}\left\{(\frac{1}{\mu_c^2} - \frac{2}{\mu_c} - 1)(2 - \frac{1}{\mu_c^2}) + \frac{1}{2\mu_c} + \frac{9}{2(2-\mu_c)}\right\}$$

and it is easy to show (study of the factor as a function of $1/\mu_c$) that $\alpha_0 + \beta_0 > 0$ for $\tau \in (0, \tau_c)$, and $\alpha_0 + \beta_0 < 0$ for $\tau > \tau_c$ where $\tau_c \in (\sqrt{6}, \sqrt{7})$, i.e. more precisely $\mu_{c,c} \sim 0.374$. We also notice that the function of μ_c in the factor for $\beta_0 - \alpha_0$ is strictly increasing for $\mu_c \in (0,1)$, and cancels for $\mu_c \sim 0.893$. Now, defining new parameters ε_1 and ε_2 by

$$A = -\frac{i\varepsilon_1}{2}, B = -\frac{i\varepsilon_2}{2},$$

for a particular choice of a solution belonging to the torus, we get the results of Theorem 2.3. □

APPENDIX C

Proof of Lemma 3.6

Let first consider the unique solution of the Cauchy problem (notice that $\frac{V_2}{V_1} \in H^{m-1}_{o,o}(\mathbb{R}^2/\Gamma) \subset C^{m-3}_{o,o}(\mathbb{R}^2/\Gamma)$, hence the vector field is Lipschitz for $m \geq 4$)

$$\frac{dX}{dx_1} = \frac{V_2}{V_1}(x_1, X), \quad X|_{x_1=0} = y,$$

which we denote by $X(x_1, y) \in C^{m-3}$. We can successively show that $X(x_1, y)$ is even in x_1, odd in y, and such that

$$X(x_1, y) = X(x_1 + 2\pi, y) = X(x_1, y + \frac{2\pi}{\tau}) - \frac{2\pi}{\tau},$$

in using the uniqueness of the solution of the Cauchy problem, in looking for the system satisfied by $X(-x_1, y)$, $-X(x_1, -y)$, $X(x_1, y + \frac{2\pi}{\tau}) - \frac{2\pi}{\tau}$, which is the same as $X(x_1, y)$, and in comparing the systems satisfied by $X(x_1 + \pi, y)$ and $X(x_1 - \pi, y)$, using the evenness in x_1. Notice that

$$\frac{\partial X}{\partial y}(x_1, y) = \exp\left(\int_0^{x_1} \partial_{x_2}(\frac{V_2}{V_1})(t, y)dt\right)$$

equals 1 for $\frac{V_2}{V_1} = 0$, i.e. if $U = 0$. Hence, for $\|U\|_4$ small enough, we can solve (implicit function theorem) with respect to $y = y(z_2)$ the equation

$$\Pi_1 X(\cdot, y) = z_2$$

where Π_1 represents the average over a period in x_1. Then, we set

$$\mathcal{Z}(Z) = X(z_1, y(z_2))$$

and, thanks to $\frac{V_2}{V_1}(x_1 + \pi, x_2 + \frac{\pi}{\tau}) = \frac{V_2}{V_1}(x_1, x_2)$ we observe that the function

$$\mathcal{Z}(z_1 + \pi, z_2 + \frac{\pi}{\tau}) - \frac{\pi}{\tau}$$

satisfies the same differential equation as $\mathcal{Z}(z_1, z_2)$ with the same average in z_1, hence by uniqueness it is identical to $\mathcal{Z}(z_1, z_2)$. It is then clear that $(z_1, z_2) \mapsto z_2 - \mathcal{Z}(Z) = \widetilde{d}_1(Z)$ satisfies the properties indicated at Lemma 3.6 hence lies in $C^{m-3}_{e,o}(\mathbb{R}^2/\Gamma)$. The proof of the tame estimates for d_1 and $\widetilde{d}_1$ is identical to the one made in Appendix G of [**IPT**]. Notice in addition, that replacing the initial condition at $x_1 = 0$ by an average condition, allows to keep the equivariance under shifts $\mathcal{T}_\delta$ parallel to x_1 direction.

APPENDIX D

Proofs of Lemmas 3.7 and 3.8

Let us start with the expressions for U, given in (2.10)
$$U = \varepsilon\xi_0 + \varepsilon^2 U^{(2)} + O(\varepsilon^3),$$
$$\mu = \mu_c + \mu_1(\tau)\varepsilon^2 + O(\varepsilon^4),$$
with $\xi_0 = (\psi_1, \eta_1)$ and $\mu_1(\tau)$ given in (2.10) and (2.11), and where

$$U^{(2)} = \begin{cases} \frac{1-2\mu_c}{4\mu_c(2-\mu_c)} \sin 2x_1 - \frac{1}{4\mu_c^2} \sin 2x_1 \cos 2\tau x_2 \\ \frac{\mu_c^2+2\mu_c-2}{4\mu_c^2(2-\mu_c)} \cos 2x_1 + \frac{\tau^2}{4\mu_c} \cos 2\tau x_2 + \frac{1}{4\mu_c^3} \cos 2x_1 \cos 2\tau x_2 \end{cases}$$
$$= (\psi_2, \eta_2).$$

We successively find from (3.1) and (3.4)
$$\mathfrak{b} = \frac{\varepsilon}{\mu_c} \sin x_1 \cos \tau x_2 + \varepsilon^2 \mathfrak{b}^{(2)} + O(\varepsilon^3),$$
$$\mathfrak{b}^{(2)} = \partial_{x_1} \eta_2 + \nabla \eta_1 \cdot \nabla \psi_1$$
$$= \left(\frac{1}{4\mu_c} + \frac{1}{2\mu_c^2} - \frac{1}{4\mu_c^3} - \frac{3}{4(2-\mu_c)} \right) \sin 2x_1 +$$
$$- \frac{1}{4\mu_c^3} \sin 2x_1 \cos 2\tau x_2,$$

$$V_1 = 1 + \varepsilon \cos x_1 \cos \tau x_2 + \varepsilon^2 V_1^{(2)} + O(\varepsilon^3),$$
$$V_2 = -\varepsilon\tau \sin x_1 \sin \tau x_2 + \varepsilon^2 V_2^{(2)} + O(\varepsilon^3),$$

$$V_1^{(2)} = \partial_{x_1} \psi_2 - \mathfrak{b}^{(1)} \partial_{x_1} \eta_1,$$
$$V_1^{(2)} = -\frac{1}{4\mu_c^2}(1 + \cos 2\tau x_2 + \cos 2x_1 \cos 2\tau x_2) +$$
$$+ \left(\frac{1}{4\mu_c} + \frac{1}{4\mu_c^2} - \frac{3}{4(2-\mu_c)} \right) \cos 2x_1,$$

$$V_2^{(2)} = \partial_{x_2} \psi_2 - \mathfrak{b}^{(1)} \partial_{x_2} \eta_1$$
$$= \frac{\tau}{4\mu_c^2} \sin 2x_1 \sin 2\tau x_2,$$

$$\mathfrak{a} = \mu_c + \frac{\varepsilon}{\mu_c} \cos x_1 \cos \tau x_2 + \varepsilon^2 \mathfrak{a}^{(2)} + O(\varepsilon^3),$$
$$\mathfrak{a}^{(2)} = \mu_1 + V^{(1)} \cdot \nabla \mathfrak{b}^{(1)} + \partial_{x_1} \mathfrak{b}^{(2)},$$

$$\mathfrak{a}^{(2)} = \mu_1 + \left(\frac{1}{\mu_c} + \frac{1}{\mu_c^2} - \frac{3}{4\mu_c^3} - \frac{3}{2(2-\mu_c)}\right)\cos 2x_1 +$$
$$+\frac{1}{4\mu_c^3}(1 - \cos 2x_1 \cos 2\tau x_2) + \frac{2\mu_c^2 - 1}{4\mu_c^3}\cos 2\tau x_2,$$

$$\frac{\{V^2 + (V \cdot \nabla \eta)^2\}^{1/2}}{V_1^3} = 1 - 2\varepsilon \cos x_1 \cos \tau x_2 + \varepsilon^2 Y^{(2)} + O(\varepsilon^3)$$

$$Y^{(2)} = \frac{3}{4\mu_c^2} + \frac{5}{8} + \left(\frac{7}{8} - \frac{1}{2\mu_c} - \frac{3}{4\mu_c^2} + \frac{3}{2(2-\mu_c)}\right)\cos 2x_1 +$$
$$+ \left(\frac{7}{8} + \frac{1}{2\mu_c^2}\right)\cos 2\tau x_2 + \left(\frac{5}{8} + \frac{1}{2\mu_c^2}\right)\cos 2x_1 \cos 2\tau x_2.$$

For the determination of the function $\mathcal{Z}(Z) = z_2 - \widetilde{d}_1(Z)$ of Lemma 3.6, we have

$$\partial_{x_1'}\widetilde{d}_1 = -\frac{V_2}{V_1}(x_1', x_2' - \widetilde{d}_1), \qquad \Pi_1 \widetilde{d}_1 = 0,$$

i.e.
$$\widetilde{d}_1 = \varepsilon \widetilde{d}_1^{(1)} + \varepsilon^2 \widetilde{d}_1^{(2)} + O(\varepsilon^3)$$

with
$$\partial_{z_1}\widetilde{d}_1^{(1)} = -V_2^{(1)}, \qquad \Pi_1 \widetilde{d}_1^{(1)} = 0,$$

$$\partial_{z_1}\widetilde{d}_1^{(2)} = V_1^{(1)} V_2^{(1)} - V_2^{(2)} + \partial_{x_2} V_2^{(1)} \widetilde{d}_1^{(1)}, \qquad \Pi_1 \widetilde{d}_1^{(2)} = 0.$$

This leads to
$$\widetilde{d}_1^{(1)} = -\tau \cos z_1 \sin \tau z_2,$$
$$\widetilde{d}_1^{(2)} = \frac{\tau}{4}\cos 2z_1 \sin 2\tau z_2,$$

hence the diffeomorphism $\mathcal{U}_1^{-1}$ takes the form

$$x_1 = z_1, \quad x_2 = z_2 + \varepsilon \tau \cos z_1 \sin \tau z_2 - \frac{\varepsilon^2 \tau}{4}\cos 2z_1 \sin 2\tau z_2 + O(\varepsilon^3).$$

Now we have
$$d_1 = \varepsilon d_1^{(1)} + \varepsilon^2 d_1^{(2)} + O(\varepsilon^3)$$

with
$$d_1^{(1)} = -\tau \cos x_1 \sin \tau x_2,$$
$$d_1^{(2)} = \frac{\tau}{4}\{\tau^2 + (1+\tau^2)\cos 2x_1\}\sin 2\tau x_2,$$

and finally (3.32) gives
$$\mathfrak{q} = \mu_c(1 - \varepsilon \cos x_1 \cos \tau x_2 + \varepsilon^2 \mathfrak{q}^{(2)}) + O(\varepsilon^3),$$

with
$$\mathfrak{q}^{(2)} = \frac{\mathfrak{a}^{(2)}}{\mu_c} + Y^{(2)} + \partial_{x_2} d_1^{(?)} - (2 + \tau^2 + \tau^1)\cos^2 x_1 \cos^2 \tau x_2$$
$$= \frac{\mu_1}{\mu_c} + \frac{1}{8} + \frac{1}{\mu_c^2} + \mathfrak{q}_{20}^{(2)} \cos 2x_1 + \mathfrak{q}_{02}^{(2)} \cos 2\tau x_2 +$$
$$+ \mathfrak{q}_{22}^{(2)} \cos 2x_1 \cos 2\tau x_2,$$
$$\mathfrak{q}_{20}^{(2)} = \frac{3}{4(2-\mu_c)} + \frac{3}{8} - \frac{5}{4\mu_c} + \frac{1}{2\mu_c^2} + \frac{1}{\mu_c^3} - \frac{1}{\mu_c^4},$$
$$\mathfrak{q}_{02}^{(2)} = \frac{7}{8} + \frac{1}{4\mu_c^2}, \quad \mathfrak{q}_{22}^{(2)} = \frac{1}{8} + \frac{1}{4\mu_c^2}.$$

Now, we obtain
$$(\mathfrak{q} \circ \mathcal{U}_1^{-1})(Z) = \mu_0 (1 - \varepsilon \cos z_1 \cos \tau z_2 + \varepsilon^2 \widetilde{\mathfrak{q}}^{(2)}) + O(\varepsilon^3)$$
with
$$\widetilde{\mathfrak{q}}^{(2)} = \frac{\mu_1}{\mu_c} - \frac{1}{8} + \frac{5}{4\mu_c^2} + (\mathfrak{q}_{20}^{(2)} + \frac{\tau^2}{4}) \cos 2z_1 + \frac{9}{8} \cos 2\tau z_2 +$$
$$+ \frac{3}{8} \cos 2z_1 \cos 2\tau z_2,$$
and
$$\{(\mathfrak{q} \circ \mathcal{U}_1^{-1})(Z)\}^{1/2} = \mu_c^{1/2} \{1 - \frac{\varepsilon}{2} \cos z_1 \cos \tau z_2 + \varepsilon^2 (\frac{\mu_1}{2\mu_c} - \frac{3}{32} + \frac{5}{8\mu_c^2}) +$$
$$+ \varepsilon^2 (\frac{\mathfrak{q}_{20}^{(2)}}{2} + \frac{\tau^2}{8} - \frac{1}{32}) \cos 2z_1 + \varepsilon^2 \frac{17}{32} \cos 2\tau z_2 +$$
$$+ \varepsilon^2 \frac{5}{32} \cos 2z_1 \cos 2\tau z_2\} + O(\varepsilon^3),$$
hence
$$\frac{1}{2\pi} \int_{-\pi}^{\pi} \{(\mathfrak{q} \circ \mathcal{U}_1^{-1})(Z)\}^{1/2} dz_1 = \mu_c^{1/2} \{1 + \varepsilon^2 (\frac{\mu_1}{2\mu_c} - \frac{3}{32} + \frac{5}{8\mu_c^2}) +$$
$$+ \varepsilon^2 \frac{17}{32} \cos 2\tau z_2\} + O(\varepsilon^3).$$

Then with (3.34), we obtain
$$\nu = \mu_c^{-1} \{1 - \varepsilon^2 (\frac{\mu_1}{\mu_c} - \frac{3}{16} + \frac{5}{4\mu_c^2}) + O(\varepsilon^3)\},$$
as indicated in Lemma 3.7. Furthermore, in using (3.33), (3.31), we notice that
$$e_2 = -\frac{17\varepsilon^2}{32\tau} \sin 2\tau z_2 + O(\varepsilon^3),$$
$$d_2 = -\frac{\varepsilon}{2} \sin z_1 \cos \tau z_2 + \varepsilon^2 (d_{21} \sin 2z_1 + d_{22} \sin 2z_1 \cos 2\tau z_2) + O(\varepsilon^3),$$
$$d_{21} = \frac{1}{4} \left\{ \frac{3}{4(2-\mu_c)} + \frac{1}{16} - \frac{5}{4\mu_c} + \frac{3}{4\mu_c^2} + \frac{1}{\mu_c^3} - \frac{1}{\mu_c^4} \right\}, \quad d_{22} = \frac{5}{64}.$$

This, with (3.27), leads to the principal part of the bi-periodic functions occurring in the diffeomorphism of the torus of Theorem 3.4:

$$d(x,y) = -\frac{\varepsilon}{2}\sin x_1 \cos \tau x_2 + \varepsilon^2 \sin 2x_1\{d_{21} - \frac{\tau^2}{8} + (d_{22} + \frac{\tau^2}{8})\cos 2\tau x_2\} + O(\varepsilon^3)$$

$$e(x,y) = -\varepsilon\tau \cos x_1 \sin \tau x_2 + \frac{\varepsilon^2}{4\tau}\sin 2\tau x_2 \left\{\frac{1}{\mu_c^4} - \frac{2}{\mu_c^2} - \frac{9}{8} + \frac{\tau^2}{\mu_c^2}\cos 2x_1\right\} + O(\varepsilon^3).$$

Inverting $\mathcal{U}_2 \circ \mathcal{U}_1$, and changing the coordinates in

$$\eta(X) = -\frac{\varepsilon}{\mu_c}\cos x_1 \cos \tau x_2 + \varepsilon^2 \left\{\frac{\mu_c^2 + 2\mu_c - 2}{4\mu_c^2(2-\mu_c)}\cos 2x_1 + \frac{\tau^2}{4\mu_c}\cos 2\tau x_2 \right.$$
$$\left. + \frac{1}{4\mu_c^3}\cos 2x_1 \cos 2\tau x_2\right\} + O(\varepsilon^3)$$

leads to the results of Lemma 3.8.

APPENDIX E

Distribution of Numbers $\{\omega_0 n^2\}$

Recall that for each $x \in \mathbb{R}$,

$$[x] = \max\{N : N \in \mathbb{N}, \quad N \leq x\}, \quad \{x\} = x - [x] \in [0, 1).$$

In this chapter we consider the distribution of the numbers

(E.1) $\qquad \theta_n = \{\omega_0 n^2 - C\}, \quad \sigma_n = \{\omega_0 n\}, \quad n \geq 1$

Due to the famous Weil Theorem [**W**], for each polynomial $f(x) = \alpha_k x^k + .. + \alpha_0$ with irrational α_k, the numbers $\{f(n)\}$ are uniformly distributed in $[0, 1]$. This means that for each $[\alpha, \beta] \subset [0, 1]$,

$$\text{card } \{n : 0 \leq n \leq N, \; \{f(n)\} \in [\alpha, \beta]\} = N(\beta - \alpha) + o(N).$$

This result is the asymptotic relation in which the remainder strongly depends on the choice of the interval. We consider the simplest case

$$f(x) = \omega_0 x^2 - C \text{ with } \{f(n)\} = \theta_n, \quad [\alpha, \beta] = [0, \rho]$$

and deduce the rough, uniform in ρ estimate which is sufficient for our needs. The main result is the following.

PROPOSITION E.1. *Suppose that*

$$\frac{1}{|e^{2\pi i \omega_0 l} - 1|} \leq c_1 l^2 \text{ for all positive integers } l.$$

Then there is a constant c depending only on c_1 such that

(E.2) $\qquad \displaystyle\frac{1}{N} \sum_{1 \leq n \leq N, \theta_n \in [0,\rho]} 1 \leq c\rho \text{ for all } \rho \in (0, 1/4) \text{ and } N > \rho^{-78}.$

The proof is based on the following lemma.

LEMMA E.2. *Under the assumptions of Proposition E.1, there exists positive c depending on c_1 only such that*

(E.3) $\qquad \displaystyle\frac{1}{N} \sum_{1 \leq n \leq N, \sigma_n \in [0,\varepsilon] \cup [1-\varepsilon,1]} 1 \leq c\varepsilon \text{ for all } \varepsilon \in (0, 1/4) \text{ and } N > \varepsilon^{-9}.$

PROOF. Fix an arbitrary positive $\varepsilon \in (0, 1/4)$ and introduce the function depending on parameter ε and given by the equalities

$$\varphi_\varepsilon(x) = 1 \text{ for } x \in [0, \varepsilon] \cup [1-\varepsilon, 1],$$
$$\varphi_\varepsilon(x) = 0 \text{ for } x \in [2\varepsilon, 1-2\varepsilon],$$
$$\varphi_\varepsilon(x) = \frac{2\varepsilon - x}{\varepsilon} \text{ for } x \in [\varepsilon, 2\varepsilon],$$
$$\varphi_\varepsilon(x) = \frac{x - 1 + 2\varepsilon}{\varepsilon} \text{ for } x \in [1-2\varepsilon, 1-\varepsilon].$$

We will assume that φ_ε is extended 1-periodically onto $\mathbb{R}$. Obviously the extended function is absolutely continuous and

(E.4) $$\int_0^1 \varphi_\varepsilon(x)dx = 3\varepsilon, \quad \int_0^1 |\varphi'_\varepsilon(x)|^2 dx = 2\varepsilon^{-1}.$$

It has the representation

$$\varphi_\varepsilon(x) = \sum_{l=-\infty}^{\infty} \varphi_{\varepsilon,l} e^{2\pi i l x}, \quad \varphi_{\varepsilon,l} = \int_0^1 e^{-2\pi i l x} \varphi_\varepsilon(x) dx.$$

It is clear that

(E.5) $$|\varphi_{\varepsilon,l}| \leq \int_0^1 \varphi_\varepsilon(x) dx = 3\varepsilon.$$

Obviously

(E.6) $$\frac{1}{N} \sum_{1 \leq n \leq N, \sigma_n \in [0,\varepsilon] \cup [1-\varepsilon,1]} 1 \leq \frac{1}{N} \sum_{1 \leq n \leq N} \varphi_\varepsilon(\sigma_n).$$

Represent φ in the form

(E.7) $$\varphi_\varepsilon(x) = \varphi_{\varepsilon,0} + \sum_{1 \leq |l| \leq k} \varphi_{\varepsilon,l} e^{2\pi i l x} + Q_k(x), \quad Q_k(x) = \sum_{k+1 \leq |l|} \varphi_{\varepsilon,l} e^{2\pi i l x}$$

We have, by equality (E.4),

$$|Q_k(x)| \leq 2 \sum_{k+1 \leq l} |\varphi_{\varepsilon,l}| \leq 2 \sqrt{\sum_{k+1 \leq l} l^{-2}} \sqrt{\sum_l l^2 |\varphi_{\varepsilon,l}|^2} =$$

(E.8) $$\frac{1}{\pi} \sqrt{\sum_{k+1 \leq l} l^{-2}} \sqrt{\int_0^1 |\varphi'_\varepsilon(x)|^2 dx} \leq c \frac{1}{\sqrt{k\varepsilon}}.$$

Here c is some absolute constant. Combining (E.6)-(E.8) we obtain

(E.9) $$\frac{1}{N} \sum_{1 \leq n \leq N, \sigma_n \in [0,\varepsilon] \cup [1-\varepsilon,1]} 1 \leq 3\varepsilon + 6\varepsilon \sum_{1 \leq l \leq k} \frac{1}{N} \left| \sum_{1 \leq n \leq N} e^{2\pi i l \sigma_n} \right| + \frac{c}{\sqrt{k\varepsilon}}.$$

Since the numbers
$$e^{2\pi i l \sigma_n} = e^{2\pi i \omega_0 l n}$$
form a geometric progression, we have
$$\left| \sum_{1 \leq n \leq N} e^{2\pi i l \sigma_n} \right| \leq \frac{2}{|e^{2\pi i \omega_0 l} - 1|} \leq 2 c_1 l^2.$$

Substituting this inequality into (E.9) we obtain
(E.10)
$$\frac{1}{N} \sum_{1 \leq n \leq N, \sigma_n \in [0,\varepsilon] \cup [1-\varepsilon,1]} 1 \leq 3\varepsilon + c \frac{\varepsilon}{N} \sum_{1 \leq l \leq k} l^2 + \frac{c}{\sqrt{k\varepsilon}} \leq 3\varepsilon + \frac{c \varepsilon k^3}{N} + \frac{c}{\sqrt{k\varepsilon}}.$$

Now set
$$k = [\varepsilon^{-3/7} N^{2/7}]$$
and note that for $N \geq \varepsilon^{-9}$,
$$\frac{1}{\sqrt{\varepsilon k}} \leq \frac{\varepsilon}{(1-\varepsilon^3)^{1/2}} \leq c\varepsilon, \quad \frac{\varepsilon k^3}{N} \leq \varepsilon,$$
which along with (E.10) yields
$$\frac{1}{N} \sum_{1 \leq n \leq N, \sigma_n \in [0,\varepsilon] \cup [1-\varepsilon,1]} 1 \leq c\varepsilon \text{ for } N \geq \varepsilon^{-9},$$
and the lemma E.2 follows. $\square$

Proof of Proposition E.1

The proof in main part imitates the proof of the Weil Theorem. Fix $\rho \in (0, 1/4)$ and consider the function φ_ρ as defined above. Obviously

(E.11) $$\frac{1}{N} \sum_{1 \leq n \leq N, \theta_n \in [0,\rho]} 1 \leq \frac{1}{N} \sum_{1 \leq n \leq N} \varphi_\rho(\theta_n).$$

Combining (E.7)-(E.8) we obtain

(E.12) $$\frac{1}{N} \sum_{1 \leq n \leq N, \theta_n \in [0,\rho]} 1 \leq 3\rho + 6\rho \sum_{1 \leq l \leq k} \frac{1}{N} \left| \sum_{1 \leq n \leq N} e^{2\pi i l \theta_n} \right| + \frac{c}{\sqrt{k\rho}}.$$

Next set

(E.13) $$W_{l,N} = \left| \sum_{1 \leq n \leq N} e^{2\pi i l \theta_n} \right|.$$

We have from (E.12)

(E.14) $$\frac{1}{N} \sum_{1 \leq n \leq N, \theta_n \in [0,\rho]} 1 \leq 3\rho + 6\rho \sum_{1 \leq l \leq k} \frac{1}{N} W_{l,N} + \frac{c}{\sqrt{k\rho}}.$$

Noting that
$$e^{2\pi i l \theta_n} = e^{2\pi i l (\omega_0 n^2 - C)}$$

we obtain

(E.15) $$W_{l,N} = \left| \sum_{1 \leq n \leq N} e^{2\pi i w_0 l n^2} \right|.$$

Next we have
$$W_{l,N}^2 = \sum_{1 \leq m,n \leq N} e^{2\pi i g w_0 l (n^2 - m^2)}.$$

Setting $r = m + n$, $q = n - m$ we arrive at
$$W_{l,N}^2 = \sum_{2 \leq r \leq N} \sum_{|q| \leq r} e^{2\pi i w_0 l r q} + \sum_{N < r \leq 2N} \sum_{|q| \leq 2N - r} e^{2\pi i w_0 l r q}$$

Introduce the quantities
$$\chi_{rl} = \left| \sum_{|q| \leq r} e^{2\pi i w_0 l r q} \right| \quad \text{for } 2 \leq r \leq N,$$

$$\chi_{rl} = \left| \sum_{|q| \leq 2N - r} e^{2\pi i w_0 l r q} \right| \quad \text{for } N < r \leq 2N$$

Thus we get

(E.16) $$W_{l,N}^2 = \sum_{2 \leq r \leq 2N} \chi_{rl}$$

Obviously

(E.17) $$\chi_{rl} \leq 2N.$$

On the other hand, since χ_{rl} is a geometric progression in q,

(E.18) $$\chi_{rl} \leq \frac{1}{\sin(\pi w_0 r l)}.$$

Choose an arbitrary $\varepsilon \in (0, 1/4)$ and denote by J_l the set of all r such that
$$2 \leq r \leq 2N, \quad \sigma_{rl} \equiv \{w_0 r l\} \in [0, \varepsilon] \cup [1 - \varepsilon, 1].$$

It is easy to see that for $r \in [2, N] \setminus J_l$,

(E.19) $$\chi_{rl} \leq \frac{1}{\sin(\pi \varepsilon)} \leq \frac{c}{\varepsilon}.$$

From this and (E.16), (E.17) we conclude that

(E.20) $$W_{l,N}^2 \leq 2N \sum_{r \in J_l} 1 + c \frac{2N}{\varepsilon}$$

On the other hand for fixed l, $r \in J_l$, and $p = rl$ we have $\sigma_p = \{w_0 p\} \in [0, \varepsilon] \cup [1 - \varepsilon, 1]$. Hence, since $l \leq k$,
$$\text{card } J_l \leq \text{card } \{p : 1 \leq p \leq 2kN, \sigma_p \in [0, \varepsilon] \cup [1 - \varepsilon, 1]\}.$$

By lemma E.2, we have
$$\frac{1}{2kN}\text{card }\{p: 1 \le p \le 2kN, \sigma_p \in [0,\varepsilon] \cup [1-\varepsilon,1]\} \le c\varepsilon \text{ for } kN \ge \varepsilon^{-9},$$
which gives
$$\sum_{r \in J_l} 1 \le c2kN\varepsilon \text{ for } kN \ge \varepsilon^{-9}.$$
Substituting this inequality in (E.20) we obtain
$$W_{l,N}^2 \le ckN^2\varepsilon + c\frac{N}{\varepsilon} \text{ for } kN > \varepsilon^{-9},$$
or
$$\frac{1}{N}W_{l,N} \le c\sqrt{k\varepsilon} + \frac{c}{\sqrt{N\varepsilon}} \text{ for } kN \ge \varepsilon^{-9}.$$
Substituting this result in (E.14) we finally obtain
$$\frac{1}{N}\sum_{1 \le n \le N, \theta_n \in [0,\rho]} 1 \le 3\rho + c\rho\sqrt{k^3\varepsilon} + \frac{c\rho k}{\sqrt{N\varepsilon}} + \frac{c}{\sqrt{k\rho}} \text{ for } kN > \varepsilon^{-9}$$
It follows from this
$$(\text{E.21}) \quad \frac{1}{N}\sum_{1 \le n \le N, \theta_n \in [0,\rho]} 1 \le 3\rho + c\rho\sqrt{k^3\varepsilon} + c\rho k^{3/2}\varepsilon^4 + \frac{c}{\sqrt{k\rho}} \text{ for } kN > \varepsilon^{-9}.$$
Now choose
$$k = [\frac{1}{\rho^3}], \quad \varepsilon = k^{-3}, \quad N \ge k^{26}.$$
Obviously
$$k > \frac{1}{2\rho^3} \ge 32, \quad \varepsilon < 1/4, \quad Nk \ge \varepsilon^{-9}, \quad k^{3/2}\varepsilon^4 \le 1, \quad k^3\varepsilon = 1.$$
From this and (E.21) we conclude that
$$\frac{1}{N}\sum_{1 \le n \le N, \theta_n \in [0,\rho]} 1 \le c\rho \text{ for } N > \rho^{-78},$$
which completes the proof of the proposition.

APPENDIX F

Pseudodifferential Operators

In this chapter we collect basic facts from the theory of pseudodifferential operators. We refer to the pioneering paper [**KN**] and monographs [**T**], [**Pe**] for general theory. Note only that different maps from functions $A(Y,k)$ to operators $\mathfrak{A} = A(Y, -i\partial_Y)$ give rise to different theories of pseudodifferential calculus. In these notes we assume that Y is a coordinate on the 2D-torus $\mathbb{T}^2$ and that the dual variable k belongs to the lattice $\mathbb{Z}^2$. The first result constitutes the continuity properties of general pseudodifferential operators.

PROPOSITION F.1. *Let $|\mathfrak{A}|^r_{0,l} < \infty$ and $0 \leq s \leq l-3$, $r+s \geq 0$. Then there is a constant c depending on s only so that for all $u \in H^{s+r}(\mathbb{R}^2/\Gamma)$,*

$$\text{(F.1)} \qquad \|\mathfrak{A}u\|_s \leq c\Big(|\mathfrak{A}|^r_{0,l}\|u\|_r + |\mathfrak{A}|^r_{0,3}\|u\|_{r+s}\Big).$$

The proof is based on the following estimate of the convolution of non-negative sequences. Let us consider a non-negative sequences $\mathbf{a}^j = (a^j(n))_{n \in \mathbb{Z}^2}$, $1 \leq j \leq m$, and $\mathbf{v} = (v(n))_{n \in \mathbb{Z}^2}$. Set

$$|\mathbf{a}^j|_s = \sup_{n \in \mathbb{Z}^2}(1+|n|)^s a^j(n), \quad |||\mathbf{v}|||_s^2 = \sum_{n \in \mathbb{Z}^2}(1+|n|)^{2s}v(n)^2.$$

LEMMA F.2. *Under the above assumptions, the convolution $\mathbf{w} = \mathbf{a}^1 * ... * \mathbf{a}^m * \mathbf{v}$ has the bound*

$$\text{(F.2)} \qquad |||\mathbf{w}|||_s \leq c(s) \sum_j \Big(\prod_{p \neq j}|\mathbf{a}^p|_3\Big)|\mathbf{a}^j|_{s+3}|||\mathbf{v}|||_0 + \Big(\prod_j |\mathbf{a}^j|_3\Big)|||\mathbf{v}|||_s.$$

PROOF. We begin with proving (F.2) for $m=1$. Recalling the formula

$$w(n) = \sum_{k_1+...+k_{m+1}=n} a^1(k_1)..a^m(k_m)v(k_{m+1})$$

and noting that for $k_1 + k_2 = n$,

$$(1+|n|)^s \leq c(s)\big((1+|k_1|)^s + (1+|k_2|)^s\big)$$

we obtain for $m=1$,

$$\begin{aligned}(1+|n|)^s w(n) &\leq c(s)|\mathbf{a}^1|_{s+3}\sum_{k_1+k_2=n}(1+|k_1|)^{-3}v(k_2) + \\ &+c(s)|\mathbf{a}^1|_3\sum_{k_1+k_2=n}(1+|k_2|)^s v(k_2)(1+|k_1|)^{-3}.\end{aligned}$$

From this and the classic inequality

$$\sum_n \left(\sum_{k_1+k_2=n} |a(k_1)v(k_2)| \right)^2 \leq \left(\sum_k |a(k)| \right)^2 \sum_k |v(k)|^2.$$

we obtain (F.2) in the case $m = 1$. The general case obviously follows from the mathematical induction principle and the distributive property of the convolution. □

Let us turn to the proof of Proposition F.1. We have

$$\widehat{\mathfrak{A}u}(n) = \frac{1}{2\pi} \sum_k \widehat{A}(n-k,k)\widehat{u}(k) \text{ where } \widehat{A}(p,k) = \frac{1}{2\pi} \int_{\mathbb{T}^2} A(Y,k) e^{-iYp} \, dY,$$

which yields

(F.3) $$|\widehat{\mathfrak{A}u}(n)| \leq c \sum_k |\widehat{A}(n-k,k)| |\widehat{u}(k)| \leq [\mathbf{a} * \mathbf{v}](n),$$

with

$$a(n) = \sup_{k \in \mathbb{Z}^2} (1+|k|)^{-r} |\widehat{A}(n,k)| \quad v(k) = (1+|k|)^r |\widehat{u}(k)|.$$

It is easy to see that

(F.4) $$|\mathbf{a}|_s \leq c |\mathfrak{A}|_{0,s}^r, \quad |||\mathbf{v}|||_s \leq c \|u\|_{s+r}.$$

Applying Lemma F.2 to (F.3), using inequalities (F.4) and noting that

$$\left(\|\mathfrak{A}u\|_s \right)^2 \leq c \sum_n (1+|n|)^{2s} |\widehat{\mathfrak{A}u}|^2$$

we obtain (F.1), and the proposition follows.

The next proposition gives the representation for the composition and commutators of pseudodifferential operators

PROPOSITION F.3. *Let $\mathfrak{A}$ and $\mathfrak{B}$ be pseudodifferential operators so that for some $r, \rho \in \mathbb{R}^1$ and non-negative integers m, l,*

$$|\mathfrak{A}|_{m,l}^r + |\mathfrak{B}|_{m,l}^\rho < \infty.$$

Let also

$$l > |r| + 5 + s, \quad l > s+3, \quad m \geq 2.$$

Then the composition $\mathfrak{AB}$ and the commutator $\mathfrak{AB} - \mathfrak{BA}$ have the representations

(F.5) $$\mathfrak{AB} = \sum_{p=0}^d (\mathfrak{AB})_p + \mathfrak{D}_{d+1}^{(AB)}, \quad d = 0, 1,$$

(F.6) $$\mathfrak{AB} - \mathfrak{BA} = \sum_{p=1}^d [\mathfrak{A}, \mathfrak{B}]_p + \mathfrak{D}_{d+1}^{[A,B]}, \quad d = 0, 1,$$

in which $(\mathfrak{A}\mathfrak{B})_p$ and $[\mathfrak{A}, \mathfrak{B}]_p$ are the pseudodifferential operators with symbols

(F.7)
$$(AB)_0(Y,k) = A(Y,k)B(Y,k), \quad (AB)_1(Y,k) = \frac{1}{i}\partial_k A(Y,k)\partial_Y B(Y,k),$$

(F.8)
$$[A,B]_0(Y,k) = 0, \quad [A,B]_1(Y,k) = \frac{1}{i}\left(\partial_k A(Y,k)\partial_Y B(Y,k) - \partial_Y A(Y,k)\partial_k B(Y,k)\right)$$
$$\text{for } k \neq 0, \text{ and } (AB)_p(Y,0) = [A,B]_p(Y,0) = 0.$$

The reminders have the estimates

$$\|\mathfrak{D}_{d+1}^{(AB)} u\|_s \leq c\left(|\mathfrak{A}|^r_{d+1,s}|\mathfrak{B}|^\rho_{d+1,|r|+d+4} + |\mathfrak{A}|^r_{d+1,3}|\mathfrak{B}|^\rho_{d+1,|r|+d+4+s}\right)\|u\|_{r+\rho-d-1} +$$
(F.9)
$$|\mathfrak{A}|^r_{d+1,3}|\mathfrak{B}|^\rho_{d+1,|r|+d+4}\|u\|_{s+r+\rho-d-1},$$

$$\|\mathfrak{D}_{d+1}^{[A,B]} u\|_s \leq c\left(|\mathfrak{A}|^r_{d+1,s+|\rho|+d+4}|\mathfrak{B}|^\rho_{d+1,|r|+d+4} +\right.$$

$$|\mathfrak{A}|^r_{d+1,|\rho|+d+4}|1 - \mathfrak{B}|^\rho_{d+1,|r|+d+4+s}\bigg)\|u\|_{r+\rho-d-1} +$$
(F.10)
$$|\mathfrak{A}|^r_{d+1,|\rho|+d+4}|1 - \mathfrak{B}|^\rho_{d+1,|r|+d+4}\|u\|_{s+r+\rho-d-1},$$

in which the constant c depends on s, r, ρ only.

The proof is based on the following lemma

LEMMA F.4. *Let*
(F.11)
$$\mathcal{R}_{d+1}(\eta,\zeta,k) = \widehat{A}(\eta,\zeta+k) - \sum_{p=0}^{d}\sum_{|\alpha|=p}\frac{1}{\alpha!(i)^{|\alpha|}}\partial_k^\alpha \widehat{A}(\eta,k)(i\zeta)^\alpha \text{ for } k \neq 0,$$

and $\mathcal{R}_{d+1}(\eta,\zeta,0) = \widehat{A}(\eta,\zeta)$, where

$$\widehat{A}(\eta,k) = \frac{1}{2\pi}\int_{\mathbb{T}^2} e^{-i\eta Y} A(Y,k) dY.$$

Then for all $\eta, \zeta, k \in \mathbb{Z}^2$ and $0 \leq s \leq l$,

(F.12) $\quad |\mathcal{R}_{d+1}(\eta,\zeta,k)| \leq c(d,r)|A|^r_{d+1,l}(1+|\eta|)^{-l}(1+|\zeta|)^{|r|+d+1}(1+|k|)^{r-d-1}.$

PROOF. It suffices to prove (4.32) for $k \neq 0$ only. If $|\zeta| \leq |k|/2$, then the Taylor formula

$$\mathcal{R}_{d+1}(\eta,\zeta,k) = \sum_{|\alpha|=d+1}\frac{d+1}{\alpha!}\left\{\int_0^1 [\partial_k^\alpha \widehat{A}](\eta,k+t\zeta)(1-t)^d\, dt\right\}\zeta^\alpha$$

implies the estimate

$$|\mathcal{R}_{d+1}(\eta,\zeta,k)| \leq c|\mathfrak{A}|^r_{d+1,s}(1+|\eta|)^{-s}(1+|\zeta|)^{d+1}\int_0^1 (1+|k+t\zeta|)^{r-d-1}dt \leq$$

(F.13) $$c|\mathfrak{A}|^r_{d+1,s}(1+|\eta|)^{-s}(1+|\zeta|)^{d+1}(1+|k|)^{r-d-1},$$

which obviously yields (4.32). If $|\zeta| \geq |k|/2$, we have

$$|\mathcal{R}_{d+1}(\zeta,k)| \leq c|\mathfrak{A}|^r_{d+1,s}(1+|\eta|)^{-s}\left[|k+\zeta|^r + \sum_0^d (1+|\zeta|)^p|k|^{r-p}\right]$$

$$\leq c|\mathfrak{A}|^r_{d+1,s}(1+|\eta|)^{-s}\left[|\zeta|^r + \sum_0^d (1+|\zeta|)^p|k|^{r-p}\right].$$

Noting that for $0 \leq p \leq d+1$ and $|\zeta| \geq |k|/2$,

$$(1+|\zeta|)^p|k|^{r-p} \leq c(1+|\zeta|)^{d+1}(1+|k|)^{r-d-1}$$

we obtain

$$|\mathcal{R}_{d+1}(\zeta,k)| \leq c(s)|\mathfrak{A}|^r_{d+1,s}(1+|\eta|)^{-s}(1+|\zeta|)^{|r|+d+1}(1+|k|)^{r-d-1},$$

and the lemma follows. $\square$

Let us turn to the proof of the proposition. Since $[\mathfrak{A},\mathfrak{B}] = -[\mathfrak{A},(1-\mathfrak{B})]$, it suffices to prove (4.33) only. To this end note that, by the definition of pseudodifferential operator,

$$\widehat{\mathfrak{A}\mathfrak{B}u}(n) = \sum_{p,k} \widehat{A}(n-p,p)\widehat{B}(p-k,k)\widehat{u}(k) = \sum_{\zeta,k} \widehat{A}(n-k-\zeta,k+\zeta)\widehat{B}(\zeta,k)\widehat{u}(k).$$

Applying Lemma F.4 to the Fourier transform $\widehat{A}(\eta,k)$ of the symbol A we arrive at the identity

$$\sum_{\substack{k,\zeta \in \mathbb{Z}^2 \\ k \neq 0}} \sum_{p=0}^d \sum_{|\alpha|=p} \frac{1}{\alpha!(i)^{|\alpha|}}[\partial_k^\alpha \widehat{A}](n-k-\zeta,k)\left[(i\zeta)^\alpha \widehat{B}(\zeta,k)\right]\widehat{u}(k)+$$

$$+ \sum_{k,\zeta \in \mathbb{Z}^2} \mathcal{R}_{d+1}(n-k-\zeta,\zeta,k)\widehat{B}(\zeta,k)\widehat{u}(k) = \widehat{\mathfrak{A}\mathfrak{B}u}(n).$$

Noting that $(i\zeta)^\alpha \widehat{B}(\zeta,k) = \widehat{\partial_Y^\alpha B}(\zeta,k)$, we obtain

$$\sum_{\substack{k,\zeta \in \mathbb{Z}^2 \\ k \neq 0}} \sum_{|\alpha|=p} \frac{1}{\alpha!(i)^{|\alpha|}}[\partial_k^\alpha \widehat{A}](n-k-\zeta)\left[(i\zeta)^\alpha \widehat{B}(\zeta,k)\right]\widehat{u}(k) = \widehat{(\mathfrak{A}\mathfrak{B})_p u}(n),$$

which leads to representation (4.33) with the remainder

$$\widehat{\mathfrak{D}_{m+1}^{(AB)}u}(n) = \sum_{k,\zeta \in \mathbb{Z}^2} \mathcal{R}_{d+1}(n-k-\zeta,\zeta,k)\widehat{B}(\zeta,k)\widehat{u}(k).$$

In particular, we have the inequality

(F.14) $$|\widehat{\mathfrak{D}_{m+1}^{(AB)}u}| \leq \mathbf{a} * \mathbf{b} * \mathbf{v},$$

in which the elements of the sequences $\mathbf{a}$, $\mathbf{b}$, $\mathbf{v}$ are given by

$$a(n) = \sup_{\zeta,k}(1+|\zeta|)^{-|r|-d-1}(1+|k|)^{-r+d+1}|\mathcal{R}_{d+1}(n,\zeta,k)|,$$
$$b(\zeta) = (1+|\zeta|)^{|r|+d+1}\sup_k(1+|k|)^{-\rho}|\widehat{B}(\zeta,k)|, \quad v(k) = (1+|k|)^{r+\rho-d-1}|\widehat{u}(k)|.$$

Applying Lemma F.2 to the right side of (F.14) and noting that by Lemma F.4

$$|\mathbf{a}|_s \leq |\mathfrak{A}|_{d+1,s}^r, \quad |\mathbf{b}|_s \leq |\mathfrak{B}|_{0,s+|r|+d+1}^\rho, \quad |||\mathbf{v}|||_s \leq c\|u\|_{\rho+r-d-1+s}$$

we obtain (4.41) and the proposition follows.

It is useful to reformulate the above results in terms of infinite matrices. To this end we introduce the Hilbert space $\mathbf{F}_{s,t}$ which consists of all operators $\mathcal{Y} : H^s(\mathbb{R}^2/\Gamma) \mapsto H^t(\mathbb{R}^2/\Gamma)$ having the representation

$$\widehat{\mathcal{Y}u}(k) = \sum_{p \in \mathbb{Z}^2} \mathcal{Y}_{kp}\widehat{u}(p)$$

such that
(F.15)
$$\|\mathcal{Y}\|_{\mathbf{F}_{s,t}}^2 := \sup\Big\{\sum_k(1+|k|)^{2t}\Big(\sum_p|\mathcal{Y}_{kp}||\widehat{u}(p)|\Big)^2 : \sum_k(1+|k|)^{2s}|\widehat{u}(k)|^2\Big\} < \infty.$$

COROLLARY F.5. *(i) Under the assumptions of Proposition F.1, operator $\mathfrak{A}$ has a matrix representation with $\mathfrak{A}_{kp} = \widehat{A}(k-p,p)$ and*

$$\sum_k(1+|k|)^{2s}\Big(\sum_p|\mathfrak{A}_{kp}||\widehat{u}(p)|\Big)^2 \leq$$
$$c(|\mathfrak{A}|_{0,l}^r)^2\sum_k(1+|k|)^{2r}|\widehat{u}(k)|^2 + c(|\mathfrak{A}|_{0,3}^r)^2\sum_k(1+|k|)^{2r+2s}|\widehat{u}(k)|^2.$$

In particular, $\|\mathfrak{A}\|_{\mathbf{F}_{r+s,s}} \leq c|\mathfrak{A}|_{0,l}^r$.

(ii) Under the assumptions of Proposition F.3, the operator $\mathfrak{D}_{d+1}^{(AB)}$ has a matrix representation so that

$$\sum_k (1+|k|)^{2(s)} \Big(\sum_p |\mathfrak{D}_{d+1,kp}^{(AB)}||\widehat{u}(p)|\Big)^2 \le c\Big(|\mathfrak{A}|_{d+1,s}^r |\mathfrak{B}|_{d+1,|r|+d+4}^\rho +$$

$$|\mathfrak{A}|_{d+1,3}^r |\mathfrak{B}|_{d+1,|r|+d+4+s}^\rho\Big)^2 \sum_k (1+|k|)^{2r+2\rho-2d-2}|\widehat{u}(k)|^2 +$$

(F.16) $\qquad c(|\mathfrak{A}|_{d+1,3}^r |\mathfrak{B}|_{d+1,|r|+d+4}^\rho)^2 \sum_k (1+|k|)^{2s+2r+2\rho-2d-2}|\widehat{u}(k)|^2$

In particular,

$$\|\mathfrak{D}_{d+1}^{(AB)}\|_{\mathbf{F}_{r+s+\rho-d-1,s}} \le c\Big(|\mathfrak{A}|_{d+1,s}^r |\mathfrak{B}|_{d+1,|r|+d+4}^\rho + |\mathfrak{A}|_{d+1,3}^r |\mathfrak{B}|_{d+1,|r|+d+4+s}^\rho\Big)$$

PROOF. Assertion (i) is integral part of the proof of Proposition F.1. In order to prove (ii) note that

$$\mathfrak{D}_{d+1,kp}^{(AB)} = \sum_{\zeta \in \mathbb{Z}^2} \mathcal{R}_{d+1}(p-k-\zeta,\zeta,k)\widehat{B}(\zeta,k),$$

where $\mathcal{R}_{d+1}$ is defined by formula (F.11). The needed result follows from (F.14) and Lemma F.2. $\qquad\square$

Proof of Proposition 5.4

We give the proof of representations (5.15) and (5.16) only, and begin with proving (5.15). Note that

$$A(Y,\xi) \equiv A_r(Y,\xi_1,\xi_2^2) + i\xi_2 A_i(Y,\xi_1,\xi_2^2),$$

where

$$A_r(\cdot,\cdot,\rho) = \frac{1}{2}\Big(A(\cdot,\cdot,\sqrt{\rho}) + A(\cdot,\cdot,-\sqrt{\rho})\Big), \quad A_i(\cdot,\cdot,\rho) = \frac{1}{2i\sqrt{\rho}}\Big(A(\cdot,\cdot,\sqrt{\rho}) - A(\cdot,\cdot,-\sqrt{\rho})\Big).$$

Assuming $k_1 \ne 0$ and noting that in this case $\xi_2^2 = 1 - \xi_1^2$, we arrive at the identity

$$A(Y,\xi) \equiv \mathcal{A}_r(Y,\xi_1) + i\xi_2 \mathcal{A}_i(Y,\xi_1) \equiv \mathcal{A}(Y,\xi) \text{ where } \mathcal{A}_\beta(Y,\xi_1) = A_\beta(Y,\xi_1, 1-\xi_1^2).$$

From the Taylor formula we conclude that for $k_1 \ne 0$,

(F.17) $\qquad \mathcal{A}(Y,\xi) = \sum_{j=0}^{j=2} \frac{1}{(i)^j j!}\big[\partial_{\xi_1}^j \mathcal{A}\big](Y,0,\xi_2)(i\xi_1)^j + (\xi_1)^3 \mathcal{R}_3(Y,\xi)$

$$\mathcal{R}_3(Y,\xi) = \frac{1}{6}\int_0^1 [\partial_{\xi_1}^3 \mathcal{A}](Y,s\xi_1,\xi_2)(1-s)^3\,ds.$$

F. PSEUDODIFFERENTIAL OPERATORS

Recalling symmetry property (5.2) we obtain

$$\mathcal{A}_r + i\xi_2\mathcal{A}_i\Big|_{\xi_1=0} = \Big[\operatorname{Re} A + i\xi_2\operatorname{Im} A\Big](Y,0,1) = A_0(Y,\xi_2),$$

$$\partial_{\xi_1}(\mathcal{A}_r + i\xi_2\mathcal{A}_i)\Big|_{\xi_1=0} = i\Big[\partial_{\xi_1}(\operatorname{Im} A - i\xi_2\operatorname{Re} A)\Big](Y,0,1) = -\nu A_1(Y,\xi_2)$$

$$\partial_{\xi_1}^2(\mathcal{A}_r + i\xi_2\mathcal{A}_i)\Big|_{\xi_1=0} = \Big[(\partial_{\xi_1}^2 - \partial_{\xi_2})(\operatorname{Re} A + i\xi_2\operatorname{Im} A) + i\xi_2\operatorname{Im} A\Big](Y,0,1) =$$
$$-2\nu^2 A_2(Y,\xi_2),$$

which being substituted into (F.17) leads to

(F.18) $\quad A(Y,\xi) = A_0(Y,\xi_2) - \nu A_1(Y,\xi_2)i\xi_1 + \nu^2 A_2(Y,\xi_2)(i\xi_1)^2 + \mathcal{R}_3(Y,\xi)\xi_1^3.$

Noting that

$$ik_1 i\xi_1 = -\frac{1}{\nu} + \frac{1}{\nu|\mathbb{T}^{-1}(k)|}L(k),$$

(F.19) $\quad ik_1(i\xi_1)^2 = \left(\frac{1}{\nu}\right)^2 \frac{1}{ik_1} - \left(\frac{1}{\nu}\right)^2 \frac{1}{ik_1|\mathbb{T}^{-1}(k)|^2}\Big(\nu k_1^2 + |\mathbb{T}^{-1}(k)|\Big)L(k),$

$$ik_1(\xi_1)^3 = \frac{i}{\nu^2|\mathbb{T}^{-1}(k)|} - \frac{i}{\nu^2|\mathbb{T}^{-1}(k)|^3}(\nu k_1^2 + |\mathbb{T}^{-1}(k)|)L(k),$$

we get for $k_1 \neq 0$,

(F.20) $\quad ik_1 A(Y,\xi) = \sum_{j=0}^{2} A_j(Y,\xi_2)(ik_1)^{1-j} + P_A(Y,\xi)L(k) + Q_A(Y,\xi).$

Here the remainders are given by

(F.21) $\quad P_A(Y,k) = \frac{1}{|\mathbb{T}^{-1}(k)|}A_1(Y,\xi_2) + \frac{i}{k_1|\mathbb{T}^{-1}(k)|^2}\Big(A_2(Y,\xi_2) -$
$\frac{k_1}{\nu^2|\mathbb{T}^{-1}(k)|}\mathcal{R}_3(Y,\xi)\Big)\Big(\nu k_1^2 + |\mathbb{T}^{-1}(k)|\Big), \quad Q_A(Y,k) = \frac{i}{\nu^2|\mathbb{T}^{-1}(k)|}\mathcal{R}_3(Y,\xi).$

Next set $P_A(Y,k) = Q_A(Y,k) = 0$ for $k_1 = 0$ and denote by $\mathfrak{P}_A, \mathfrak{Q}_A$ the pseudodifferential operators with the symbols P_A, Q_A. With this notation, decomposition (5.15) easy follows from (F.20). It remains to note that formulae (F.17) and (F.21) imply the estimate

$$|\mathfrak{P}_A|_{0,l}^{-1} + |\mathfrak{Q}_A|_{0,l}^{-1} \leq |\mathfrak{A}|_{3,l},$$

which along with Proposition F.1 yields inequalities (5.20) for $\mathfrak{P}_A$ and $\mathfrak{Q}_A$.

Let us turn to the proof of (5.16). We begin with the observation that for $k_1 \neq 0$, the product of the symbols of the elementary operators $\mathfrak{A}_j$ and $\mathfrak{W}$ is equal to

(F.22) $\qquad A_j W \equiv S_j' + \xi_1^2 \operatorname{Im} \tilde{A}_j \operatorname{Im} \tilde{W},$

where S_j' are the symbols of the elementary operators associated with the functions $\tilde{A}_j \tilde{W}$. Multiplying both sides of (F.20) by $W(Y,\xi_2)$ and using the

identities
$$ik_1\xi_1^2 = -\left(\frac{1}{\nu}\right)^2 \frac{1}{ik_1} + \left(\frac{1}{\nu}\right)^2 \frac{1}{|\mathbb{T}^{-1}(k)|ik_1}\left(\nu k_1^2 + |\mathbb{T}^{-1}(k)|\right)L(k),$$
$$\xi_1^2 = \frac{1}{\nu}\frac{1}{|\mathbb{T}^{-1}(k)|} - \frac{1}{\nu}\frac{1}{|\mathbb{T}^{-1}(k)|^2}L(k)$$

we arrive at

$$ik_1 S(Y,k) = \sum_{j=0}^{2}(ik_1)^{1-j} S_j'(Y,\xi_2) - \left(\frac{1}{\nu}\right)^2 \operatorname{Im} \tilde{A}_0 \operatorname{Im} \tilde{W} \frac{1}{ik_1} + L(k)U_S(Y,k) + V_S(Y,k),$$

where the symbols $U_S(Y,k)$ and $V_S(Y,k)$ are defined by

$$U_S = P_A W + \left(\frac{1}{\nu}\right)^2 \frac{1}{|\mathbb{T}^{-1}(k)|ik_1}\left(\nu k_1^2 + |\mathbb{T}^{-1}(k)|\right)\operatorname{Im} \tilde{A}_0 \operatorname{Im} \tilde{W} -$$
$$\frac{1}{\nu|\mathbb{T}^{-1}(k)|^2}\left(\operatorname{Im} \tilde{A}_1 + (ik_1)^{-1}\operatorname{Im} \tilde{A}_2\right)\operatorname{Im} \tilde{W},$$
$$V_S = Q_A W + \frac{1}{\nu|\mathbb{T}^{-1}(k)|}\left(\operatorname{Im} \tilde{A}_1 + (ik_1)^{-1}\operatorname{Im} \tilde{A}_2\right)\operatorname{Im} \tilde{W}.$$

Recall that $\tilde{W}$ and $\tilde{A}$ are smooth function on tori $\mathbb{T}^2$ which do not depend on k. Setting $U_S(Y,k) = V_S(Y,k) = 0$ for $k_1 = 0$, denoting by $\mathfrak{U}_S$ and $\mathfrak{V}_S$ the pseudodifferential operators with the symbols U_S, V_S and arguing as before we obtain desired identity (5.16). It remains to note that estimate (5.21) follows from Proposition (F.1) which completes the proof.

APPENDIX G

Dirichlet-Neumann Operator

In this chapter we deduce the basic decomposition for the Dirichlet-Neumann operator and prove Theorem 3.5. Let us denote the change of coordinates by

(G.1) $$\begin{aligned} x &= x(y), \quad x = (X, x_3), \quad y = (Y, y_3), \\ X &= X(Y), \quad x_3 = y_3 + \tilde{\eta}(Y), \end{aligned}$$

where
$$\tilde{\eta}(Y) = \eta(X(Y)),$$

and $X(\cdot)$ is a diffeomorphism of the torus of the form (3.8) satisfying the Condition 3.2. Let us notice that with the new coordinate $Y = (y_1, y_2)$ the lattice Γ' is generated by the two wave vectors $(1, \pm 1)$. We still denote the lattice of periods by Γ. For a function $u(x)$ we define $\tilde{u}(y)$ by $\tilde{u}(y) = u(x(y))$. The Jacobian matrix of the above diffeomorphism reads

$$\mathbb{B}_1(Y) = \begin{pmatrix} \partial_{y_1} X_1 & \partial_{y_2} X_1 & 0 \\ \partial_{y_1} X_2 & \partial_{y_2} X_2 & 0 \\ \partial_{y_1} \tilde{\eta} & \partial_{y_2} \tilde{\eta} & 1 \end{pmatrix}$$

and the determinant satisfies $J = \det \mathbb{B}(Y) = \det \mathbb{B}_1(Y)$. Now, we use the following identities for any scalar function u, and vector function V:

$$\begin{aligned} \nabla_x u(x(y)) &= (\mathbb{B}_1^*)^{-1} \nabla_y \tilde{u}(y), \\ \nabla_x \cdot V(x(y)) &= \frac{1}{J} \nabla_y \cdot (J \mathbb{B}_1^{-1} \tilde{V}(y)). \end{aligned}$$

With these identities, the Dirichlet-Neumann operator (1.4) takes the new following form

$$\begin{aligned} \mathcal{A}\tilde{\varphi} &= 0, \quad y_3 \in (-\infty, 0), \\ \tilde{\varphi}|_{y_3=0} &= \tilde{\psi}(Y), \\ \nabla \tilde{\varphi} &\to 0 \text{ as } y_3 \to -\infty, \end{aligned}$$

(G.2) $$\mathcal{G}_\eta \psi = \frac{1}{J}(\mathbb{A}(Y) \nabla_y \tilde{\varphi}) \cdot e_3,$$

where
$$\begin{aligned} \mathcal{A}\tilde{\varphi} &= \nabla_y \cdot (\mathbb{A}(Y) \nabla_y \tilde{\varphi}), \\ \mathbb{A} &= J(\mathbb{B}_1^* \mathbb{B}_1)^{-1} \text{ (symmetric matrix)} \\ \det \mathbb{A} &= J. \end{aligned}$$

Notice that for computing the new expression of $\mathcal{G}_\eta \psi$, we used the fact that $\Phi(x) = x_3 - \eta(X)$ is such that $\widetilde{\Phi}(y) = y_3$, hence $\nabla_x \Phi = (\mathbb{B}_1^*)^{-1} \mathbf{e}_3$.

We already defined the 2x2 matrix $\mathbb{G}(Y)$ of the first fundamental form of the free surface, and we have

$$\mathbb{B}_1^* \mathbb{B}_1(Y) = \begin{pmatrix} g_{11} & g_{12} & \partial_{y_1} \widetilde{\eta} \\ g_{12} & g_{22} & \partial_{y_2} \widetilde{\eta} \\ \partial_{y_1} \widetilde{\eta} & \partial_{y_2} \widetilde{\eta} & 1 \end{pmatrix}.$$

We assume that elements of the matrix $\mathbb{A}(Y)$ are smooth 2π-periodic functions, and for some ρ and l, satisfying the following inequalities for $9 \leq \rho \leq l$,

(G.3) $\quad \|\mathbb{A}_0 - \mathbb{A}\|_{C^\rho} \leq c\varepsilon, \quad \|\mathbb{A}\|_{C^l} \leq E_l$, where $\mathbb{A}_0 = \tau^{-1} \text{diag}\{1, \tau^2, 1\}$,

where, by construction, the estimates of $\mathbb{A}$ and $\mathbb{A}_0 - \mathbb{A}$ would come from

(G.4) $\quad \begin{aligned} \|\widetilde{\eta}\|_{C^{\rho+1}} + \|\widetilde{\mathcal{V}}(\cdot)\|_{C^{\rho+1}} &\leq c\varepsilon, \\ \|\widetilde{\eta}\|_{C^{s+1}} + \|\widetilde{\mathcal{V}}(\cdot)\|_{C^{s+1}} &\leq c(l), \quad s \leq l, \end{aligned}$

where $\widetilde{\mathcal{V}}$ is defined in (3.8). By the factorization theorem, there are first order pseudodifferential operators $\mathcal{G}^\pm$ so that

(G.5) $\quad \mathcal{A} = a_{33}(\partial_{y_3} + \mathcal{G}^+)(\partial_{y_3} + \mathcal{G}^-),$
$\quad e^{-ik\cdot Y} \mathcal{G}^\pm e^{ik\cdot Y} \to \pm\infty$ as $|k| \to \infty$.

It follows from (G.2) that

(G.6) $\quad \mathcal{G}_\eta \psi = \frac{1}{J}\Big(\sum_{j=1}^{2} a_{3j} \partial_{y_j} - a_{33} \mathcal{G}^-\Big) \widetilde{\psi}.$

Hence the task now is to split $\mathcal{G}^-$ into a sum of first and zero order pseudodifferential operators. The corresponding result is given by the following

THEOREM G.1. *Under the above assumptions there is ε_0 depending on ρ and l only such that for all $\varepsilon \in (0, \varepsilon_0)$, the operator $\mathcal{G}^-$ has the representation*

(G.7) $\quad \mathcal{G}^- = \mathcal{G}_1^- + \mathcal{G}_0^- + \mathcal{G}_{-1}^-,$

in which the pseudodifferential operators $\mathcal{G}_0^-, \mathcal{G}_1^-$, have the symbols G_0^- and G_1^- defined by

(G.8) $\quad G_0^- = \frac{a_{33}}{2D}\big(i\nabla_k G_1^+ \nabla_Y G_1^- - 2bG_1^- + C_1\big),$

(G.9) $\quad G_1^\pm(Y, k) = \frac{1}{a_{33}}(ia_{31}k_1 + ia_{32}k_2 \pm D),$

where

(G.10) $\quad D(Y, k) = \Big\{\sum_{1 \leq j,m \leq 2} a_{33} a_{jm} k_j k_m - (a_{31}k_1 + a_{32}k_2)^2\Big\}^{1/2},$

$\quad C_1(Y, k) = \frac{i}{a_{33}} \sum_{j,m=1}^{2} (\partial_{y_j} a_{jm}) k_m, \quad 2b = \frac{1}{a_{33}} \sum_{j=1}^{2} \partial_{y_j} a_{3j}.$

The zero-order pseudodifferential operator $\mathcal{G}_0^-$ satisfies the inequality

(G.11) $\qquad |\mathcal{G}_0^-|_{m,n-1}^0 \leq c\|\mathbb{A} - \mathbb{A}_0\|_{C^n}, \quad m \geq 0, \quad n \leq l,$

and for any $u \in H^{r-1}(\mathbb{R}^2/\Gamma)$, $r < \rho - 8$ and $1 \leq s \leq l-8$, the rest term has the bound

(G.12) $\qquad \|\mathcal{G}_{-1}^- u\|_r \leq c\varepsilon \|u\|_{r-1}, \quad \|\mathcal{G}_{-1}^- u\|_s \leq c(E_l \|u\|_{r-1} + \varepsilon \|u\|_{s-1}).$

Moreover, operators $\mathcal{G}_1^-$, $\mathcal{G}_0^-$, $\mathcal{G}_{-1}^-$ verify the following symmetry properties

(G.13) $\qquad \mathcal{G}_j^- u(\pm Y^*) = \mathcal{G}_j^- u^*(\pm Y), \quad j = 1, 0, -1, \quad u^*(Y) = u(Y^*).$

PROOF. First we rewrite the operator $\mathcal{A}$ in the form

(G.14) $\qquad \mathcal{A} = a_{33}\dfrac{\partial^2}{\partial y_3^2} + 2a_{33}\dfrac{\partial}{\partial y_3}\mathcal{B} + a_{33}\mathcal{C}, \quad \mathcal{C} = \mathcal{C}_2 + \mathcal{C}_1, \quad \mathcal{B} = \mathcal{B}_1 + b$

with differential operators

(G.15)
$$\mathcal{C}_2 = a_{33}^{-1} \sum_{i,j=1}^{2} a_{ij} \partial_{y_i} \partial_{y_j}, \quad \mathcal{C}_1 = a_{33}^{-1} \sum_{i,j=1}^{2} (\partial_{y_i} a_{ij}) \partial_{y_j},$$
$$\mathcal{B}_1 = a_{33}^{-1} \sum_{j=1}^{2} a_{3j} \partial_{y_j}, \quad 2b = a_{33}^{-1} \sum_{j=1}^{2} \partial_{y_j} a_{3j}.$$

Combining (G.5) and (G.14) we obtain

(G.16) $\qquad \mathcal{G}^+ + \mathcal{G}^- = 2\mathcal{B}, \quad \mathcal{G}^+ \mathcal{G}^- = \mathcal{C}.$

We find the solution of (G.16) in the form

(G.17) $\qquad \mathcal{G}^\pm = \mathcal{G}_1^\pm + \mathcal{G}_0^\pm \pm \mathcal{X},$

with

(G.18) $\qquad \mathcal{G}_1^+ + \mathcal{G}_1^- = 2\mathcal{B}_1, \quad \mathcal{G}_0^+ + \mathcal{G}_0^- = 2b,$

where the symbol of pseudodifferential operator $\mathcal{G}_1^-$ is given by formula (G.9), $\mathcal{G}_0^-$ satisfying (G.11) and $\mathcal{X}$ being unknown. It follows from these formulae that
$$\mathcal{G}_1^\pm = \pm(-\Delta)^{1/2} + \tilde{\mathcal{G}}_1^\pm$$
and for any integers $m \geq 0$ and $n \leq l$

(G.19) $\qquad |\tilde{\mathcal{G}}_1^-|_{m,n}^1 + |\mathcal{G}_0^-|_{m,n-1}^0 + |\mathcal{B}|_{m,n-1}^1 \leq c\|\mathbb{A} - \mathbb{A}_0\|_{C^n}.$

Representation (F.5) from Proposition F.3 yields the decompositions

(G.20)
$$\mathcal{G}_1^+ \mathcal{G}_1^- = \mathcal{G}_1^{(2)} + \mathcal{G}_1^{(1)} + \mathcal{R}_1,$$
$$\mathcal{G}_0^+ \mathcal{G}_1^- = \mathcal{G}_{01}^{(1)} + \mathcal{R}_{01}, \quad \mathcal{G}_1^+ \mathcal{G}_0^- = \mathcal{G}_{10}^{(1)} + \mathcal{R}_{10}.$$

Here the second order pseudodifferential operator $\mathcal{G}_1^{(2)}$ have the symbol $G_1^+ G_1^-$, and satisfies the identity

(G.21) $\qquad \mathcal{G}_1^{(2)} = \mathcal{C}_2,$

the first order pseudodifferential operators $\mathcal{G}_1^{(1)}$, $\mathcal{G}_{10}^{(1)}$, and $\mathcal{G}_{01}^{(1)}$ have the symbols

(G.22)
$$G_1^{(1)} = -i\partial_k G_1^+ \partial_Y G_1^-,$$
$$G_{10}^{(1)} = G_1^+ G_0^-, \quad G_{01}^{(1)} = G_0^+ G_1^-,$$

and satisfy the identity

(G.23) $$\mathcal{G}_1^{(1)} + \mathcal{G}_{10}^{(1)} + \mathcal{G}_{01}^{(1)} - \mathcal{C}_1 = 0.$$

Now from (G.16) and (G.21) we have

$$G_1^+ + G_1^- = 2B_1, \quad G_1^+ G_1^- = C_2$$

which leads to

$$G_1^{\pm} = B_1 \pm (B_1^2 - C_2)^{1/2}$$

and since $a_{33} > 0$ for ε small enough, and noticing from (G.10) that

$$(B_1^2 - C_2)^{1/2} = \frac{1}{a_{33}} D,$$

the formula (G.9) follows. For obtaining G_0^-, we use (G.23), (G.22) which leads to (G.8)

It is then clear that G_0^- is a pseudodifferential operator of zero order which satisfies (G.11). Now, the symmetry properties (G.13) for $\mathcal{G}_1^-$ and $\mathcal{G}_0^-$ follow from (G.9), (G.8), the evenness of a, a_{33}, a_{11}, a_{22} in y_1 and y_2, the oddness of b and a_{12} in y_1 and y_2, and the evenness in y_1, oddness in y_2 of a_{13}, and the oddness in y_1, evenness in y_2 of a_{23}.

Inequality (F.9) from Proposition F.3 along with (G.19) implies the estimates

(G.24)
$$\|\mathcal{R}_{01}u\|_s + \|\mathcal{R}_{10}u\|_s + \|\mathcal{R}_1 u\|_s \leq c\|\mathbb{A} - \mathbb{A}_0\|_{C^{6+s}} \|\mathbb{A} - \mathbb{A}_0\|_{C^6} \|u\|_0 +$$
$$+ c\|\mathbb{A} - \mathbb{A}_0\|_{C^6}^2 \|u\|_s.$$

Substituting identities (G.20), (G.21) and (G.23) into (G.16) gives the equation for the operator $\mathcal{X}$

(G.25) $$\mathcal{X}^2 + (-\Delta)^{1/2} \mathcal{X} + \mathcal{X}(-\Delta)^{1/2} + \mathcal{U}\mathcal{X} + \mathcal{X}\mathcal{V} + \mathcal{W} = 0,$$

where

$$\mathcal{U} = (\widetilde{\mathcal{G}}_1^+ + \mathcal{G}_0^+), \mathcal{V} = -(\widetilde{\mathcal{G}}_1^- + \mathcal{G}_0^-),$$
$$\mathcal{W} = -(\mathcal{R}_1 + \mathcal{R}_{10} + \mathcal{R}_{01} + \mathcal{G}_0^+ \mathcal{G}_0^-).$$

Our task is to prove the existence of a "small" solution $\mathcal{X}$ to (G.25).

Introduce the Banach spaces of bounded operators

$$\mathbf{X}_s = \mathbf{F}_{s-1,s} \cap \mathbf{F}_{s,s+1}, \quad \mathbf{Y}_s = \mathbf{F}_{s,s-1}, \quad \mathbf{Z}_s = \mathbf{F}_{s,s}$$

supplemented with the norms

$$\|\mathcal{X}\|_{\mathbf{X}_s} = \|\mathcal{X}\|_{\mathbf{F}_{s,s+1}} + \|\mathcal{X}\|_{\mathbf{F}_{s-1,s}}, \quad \|\mathcal{Y}\|_{\mathbf{Y}_s} = \|\mathcal{Y}\|_{\mathbf{F}_{s,s-1}}, \quad \|\mathcal{Z}\|_{\mathbf{Z}_s} = \|\mathcal{Z}\|_{\mathbf{F}_{s,s}}.$$

It follows from (G.19), Corollary F.5, and (G.24) that for all $r \leq \rho - 8$,

(G.26) $$\|\mathcal{U}\|_{\mathbf{Y}_{r+1}} + \|\mathcal{V}\|_{\mathbf{Y}_r} + \|\mathcal{W}\|_{\mathbf{Z}_s} \leq c\varepsilon,$$

where the constant c depends on l and ρ only. The rest of the proof is based on the following lemma which is proved at the end of this chapter.

LEMMA G.2. *Under the above assumptions, there exist $\varepsilon_0 > 0$ and $c > 0$ depending on ρ and l only such that for $\varepsilon \in (0, \varepsilon_0)$ equation (G.25) has a solution satisfying the inequalities*

(G.27) $$\|\mathcal{X}\|_{\mathbf{X}_r} \leq c\varepsilon \text{ when } 0 \leq r \leq \rho - 8,$$
$$\|\mathcal{X}u\|_s \leq c(\varepsilon \|u\|_{s-1} + E_l \|u\|_0) \text{ when } 1 \leq s \leq l - 8.$$

Moreover the operator $\mathcal{X}$ satisfies the following symmetry property

$$\mathcal{X}u(\pm Y^*) = \mathcal{X}u^*(\pm Y).$$

PROOF. We start with the consideration of the simple linear operator equation

(G.28) $$(-\Delta)^{1/2}\mathcal{X} + \mathcal{X}(-\Delta)^{1/2} = \mathcal{Z} \text{ with } \mathcal{Z} \in \mathbf{Z}_s.$$

It is easy to see that for any $\mathcal{Z} \in \mathbf{Z}_s$ satisfying (G.28) and having the matrix form with elements $\mathcal{Z}_{kp}$, operator $\mathcal{X}$ has the matrix representation with elements

$$\mathcal{X}_{kp} = \frac{1}{N(k) + N(p)} \mathcal{Z}_{kp}, \text{ where } N(k) = (k_1^2 + \tau^2 k_2^2)^{1/2}$$

which obviously yields the estimate

(G.29) $$\|\mathcal{X}\|_{\mathbf{X}_s} \leq c(\tau)\|\mathcal{Z}\|_{\mathbf{Z}_s}.$$

Hence the mapping $\mathcal{Z} \mapsto \mathcal{X}$ defines a bounded linear operator $\Xi \in \mathcal{L}(\mathbf{Z}_s, \mathbf{X}_s)$. Let us consider the sequence of operators $\mathcal{X}_n$ defined by the equalities

$$\mathcal{X}_0 = 0, \quad \mathcal{X}_{n+1} = -\Xi\left(\mathcal{X}_n^2 + \mathcal{U}\mathcal{X}_n + \mathcal{X}_n\mathcal{V} + \mathcal{W}\right).$$

Note that $\mathcal{U}$ is a pseudodifferential operator which symbol $U(Y, k)$ satisfying the inequalities

$$\|U(\cdot, k)\|_{C^r} \leq c\varepsilon, \quad |\widehat{U}(p, q)| \leq c\varepsilon(1 + |p|)^{-\rho+1}|q|.$$

It is easy to see that for any $\mathcal{X} \in \mathbf{X}_r$, the operator $\mathcal{U}\mathcal{X}$ has a matrix representation with the matrix elements

$$(\mathcal{U}\mathcal{X})_{kp} = \sum_{q \in \mathbb{Z}^2} \widehat{U}(k - q, q)\mathcal{X}_{qp}.$$

We have

$$(1 + |k|)^r |(\mathcal{U}\mathcal{X})_{kp}| \leq c\varepsilon \sum_q (1 + |k - q|)^{r-\rho+1} |\mathcal{X}_{qp}|(1 + |q|)^{r+1},$$

which gives
$$\sum_k (1+|k|)^{2r}\Big[\sum_p |(\mathcal{U}\mathcal{X})_{kp}||\widehat{u}(p)|\Big]^2 \le$$
$$c\varepsilon^2 \sum_k \Big[\sum_p \sum_q (1+|q-k|)^{r-\rho+1}(1+|q|)^{r+1}|\mathcal{X}_{qp}||\widehat{u}(p)|\Big]^2 \le$$
$$c\varepsilon^2 \sum_k \Big[\sum_q (1+|q-k|)^{r-\rho+1}\Big((1+|q|)^{r+1}\sum_p |\mathcal{X}_{qp}||\widehat{u}(p)|\Big)\Big]^2 \le$$
$$c\varepsilon^2 \sum_q \Big[(1+|q|)^{r+1}\sum_p |\mathcal{X}_{qp}||\widehat{u}(p)|\Big]^2 \le c\varepsilon^2 \|\mathcal{X}\|^2_{\mathbf{F}_{r,r+1}} \sum_p (1+|p|)^{2r}|\widehat{u}(p)|^2.$$

Thus we get
$$\|\mathcal{U}\mathcal{X}\|_{\mathbf{Z}_r} \le c\|\mathcal{U}\|_{\mathbf{Y}_{r+1}}\|\mathcal{X}\|_{\mathbf{X}_r}.$$
Repeating these arguments we obtain
$$\|\mathcal{X}\mathcal{V}\|_{\mathbf{Z}_r} \le c\|\mathcal{V}\|_{\mathbf{Y}_r}\|\mathcal{X}\|_{\mathbf{X}_r},$$
and
$$\|\mathcal{X}^2\|_{\mathbf{Z}_r} \le c\|\mathcal{X}\|^2_{\mathbf{Z}_r} \le c\|\mathcal{X}\|^2_{\mathbf{X}_r},$$
inequality (G.29) yields the estimates

(G.30) $\qquad \|\mathcal{X}_{n+1}\|_{\mathbf{X}_r} \le c\varepsilon(\|\mathcal{X}_n\|_{\mathbf{X}_r}+1) + \|\mathcal{X}_n\|^2_{\mathbf{X}_r},$

which holds true for all $r \in (1, \rho - 8)$. On the other hand, since
$$\mathcal{X}^2_{n+1} - \mathcal{X}^2_n = (\mathcal{X}_{n+1} - \mathcal{X}_n)\mathcal{X}_{n+1} + \mathcal{X}_n(\mathcal{X}_{n+1} - \mathcal{X}_n)$$
we have

(G.31) $\qquad \|\mathcal{X}_{n+1} - \mathcal{X}_n\|_{\mathbf{X}_r} \le c\|\mathcal{X}_n - \mathcal{X}_{n-1}\|_{\mathbf{X}_r}(\varepsilon + \|\mathcal{X}_n\|_{\mathbf{X}_r} + \|\mathcal{X}_{n-1}\|_{\mathbf{X}_r}).$

Here the constant c depends on ϱ and τ only. It follows from (G.30) that for all $\varepsilon \in (0, \varepsilon_0(\rho, \tau))$, the values $\|\mathcal{X}_n\|_{\mathbf{X}_r}$ are less than $c\varepsilon$. From this and (G.31) we conclude that for all small ε the sequence $\mathcal{X}_n$ converges in $\mathbf{X}_r$. Repeating these arguments and using Corollary F.5 gives the tame estimate (G.27), and the lemma follows once we observe that the symmetry property of $\mathcal{X}$ follows from the uniqueness of $\mathcal{X}$ and from the equivariance with respect to the required symmetry of the equation (G.25). $\qquad \square$

In order to complete the proof of Theorem G.1, it remains to note that operator $\mathcal{G}^-_{-1} := -\mathcal{X}$ with $\mathcal{X}$ given by Lemma G.2 satisfies (G.12). $\qquad \square$

Proof of Theorem 3.5 It follows from formulae (G.9), (G.6), and (G.7) that we can write
$$\mathcal{G}_\eta \psi = \frac{1}{J}\{\mathcal{D} - a_{33}(\mathcal{G}^-_0 + \mathcal{G}^-_{-1})\}\widetilde{\psi},$$
where the first order pseudodifferential operator $\mathcal{D}$ has the symbol $D(Y, k)$. Then we define
$$\mathcal{G}_0 = -\frac{a_{33}}{J}\mathcal{G}^-_0, \quad \mathcal{G}_{-1} = -\frac{a_{33}}{J}\mathcal{G}^-_{-1},$$

and the symmetry properties follow from the evenness of J, a_{33} and from Theorem G.1 and Lemma G.2. The zero order pseudodifferential operator $\mathcal{G}_0$ satisfies (3.22) and from Proposition F.1 we have

$$
\begin{aligned}
\|\mathcal{G}_0 u\|_r &\leq c\varepsilon \|u\|_r, \quad 0 \leq r \leq \rho - 4 \\
\|\mathcal{G}_0 u\|_s &\leq c(\varepsilon \|u\|_s + E_l \|u\|_0), \quad 0 \leq s \leq l - 4,
\end{aligned}
\tag{G.32}
$$

while the operator $\mathcal{G}_{-1}$ satisfies

$$
\begin{aligned}
\|\mathcal{G}_{-1} u\|_r &\leq c\varepsilon \|u\|_{r-1}, \quad 1 \leq r \leq \rho - 8, \\
\|\mathcal{G}_{-1} u\|_s &\leq c(\varepsilon \|u\|_{s-1} + E_l \|u\|_0), \quad 1 \leq s \leq l - 8.
\end{aligned}
\tag{G.33}
$$

We then deduce the estimates (3.23) in using (G.4).

Now our task is to calculate the symbols G_j. It is convenient to introduce the scalar $\mathbf{I}(Y)$ and linear form $\Pi(Y, k)$ defined by

$$
\mathbf{I} = J/\sqrt{\det \mathbb{G}}, \quad \Pi = \mathbb{G}^{-1} \nabla_Y \tilde{\eta} \cdot k.
$$

Recall the identity

$$
a_{33} \sum_{1 \leq j,m \leq 2} a_{jm} k_j k_m - \left(\sum_{1 \leq m \leq 2} a_{3m} k_m \right)^2
$$
$$
= \det \mathbb{A} \left((\mathbb{A}^{-1})_{22} k_1^2 - 2(\mathbb{A}^{-1})_{12} k_1 \xi_2 + (\mathbb{A}^{-1})_{11} k_2^2 \right).
$$

Noting that $\det \mathbb{A} = J$, $\mathbb{A}^{-1} = J^{-1} \mathbb{B}_1^* \mathbb{B}_1$, we conclude from this that

$$
a_{33} \sum_{1 \leq j,m \leq 2} a_{jm} k_j k_m - \left(\sum_{1 \leq m \leq 2} a_{3m} k_m \right)^2 = (\mathbb{B}_1^* \mathbb{B}_1)_{22} k_1^2 - 2(\mathbb{B}_1^* \mathbb{B}_1)_{12} k_1 k_2 + (\mathbb{B}_1^* \mathbb{B}_1)_{11} k_2^2.
$$

Hence, by the definition of the metric tensor $\mathbb{G}$,

$$
a_{33} \sum_{1 \leq j,m \leq 2} a_{jm} k_j \xi_m - \left(\sum_{1 \leq m \leq 2} a_{3m} k_m \right)^2 = g_{22} k_1^2 - 2g_{12} k_1 k_2 + g_{11} k_2^2,
$$

which yields

$$
a_{33} \sum_{1 \leq j,m \leq 2} a_{jm} k_j k_m - \left(\sum_{1 \leq m \leq 2} a_{3m} k_m \right)^2 = (\det \mathbb{G}) \mathbb{G}^{-1} k \cdot k.
\tag{G.34}
$$

Substituting this relation into D (see (G.10)) finally gives

$$
D(Y, k) = \sqrt{\det \mathbb{G}}\, \mathbf{G}_1(Y, k).
\tag{G.35}
$$

Noting that $\mathcal{G}_1 = \frac{1}{J} \mathcal{D}$ we obtain the needed formula (3.18). The calculation of G_0 is more delicate task. Since $a_{33} = \det \mathbb{G}/J$, formula (G.8) yields

$$
(2 \mathbf{I} \mathbf{G}_1) G_0^- = i \nabla_k G_1^+ \nabla_Y G_1^- - \frac{G_1^-}{a_{33}} \sum_{j=1}^{2} \partial_{y_j} a_{3j} + \frac{i}{a_{33}} \sum_{j,m=1}^{2} \partial_{y_j} a_{jm} k_m.
$$

It follows from the definition of the form Π that

$$
\sum_{m=1}^{2} a_{3m} k_m = -a_{33} \Pi, \quad \sum_{j=1}^{2} \partial_{y_j} a_{3j} = - \operatorname{div}_Y (a_{33} \nabla_k \Pi).
\tag{G.36}
$$

From this, (G.9), and (G.35) we conclude that
$$G_1^- = -\mathbf{IG}_1 - i\Pi, \quad G_1^+ = \mathbf{IG}_1 - i\Pi,$$
and hence
(G.37)
$$(2\mathbf{IG}_1)G_0^- = i\nabla_k(\mathbf{IG}_1-i\Pi)\nabla_Y(-\mathbf{IG}_1-i\Pi) - \frac{1}{a_{33}}(\mathbf{IG}_1+i\Pi)\,\mathrm{div}_Y(a_{33}\nabla_k\Pi) +$$
$$\frac{i}{a_{33}}\sum_{j,m=1}^{2}\partial_{y_j}a_{jm}k_m.$$

Next differentiating both sides of (G.34) with respect to k_j we arrive to
$$2a_{33}\sum_{m=1}^{2}a_{jm}k_m - 2a_{3j}\sum_{m=1}^{2}a_{3m}k_m = (\det\mathbb{G})\partial_{k_j}\mathbf{G}_1^2,$$
which along with (G.36) and the identity $\det\mathbb{G}/a_{33} = J$ leads to
$$\sum_{m=1}^{2}a_{jm}k_m = a_{33}\Pi\partial_{k_j}\Pi + J\mathbf{G}_1\partial_{k_j}\mathbf{G}_1.$$
Substituting this expression into (G.37) we finally obtain
(G.38)
$$(2\mathbf{IG}_1)G_0^- = i\nabla_k(\mathbf{IG}_1-i\Pi)\nabla_Y(-\mathbf{IG}_1-i\Pi) - \frac{1}{a_{33}}(\mathbf{IG}_1+i\Pi)\,\mathrm{div}_Y(a_{33}\nabla_k\Pi) +$$
$$\frac{i}{a_{33}}\mathrm{div}_Y\Big(a_{33}\Pi\nabla_k\Pi + J\mathbf{G}_1\nabla_k\mathbf{G}_1\Big).$$
Let us calculate the real part of G_0. It is easy to see that
$$(2\mathbf{IG}_1)\,\mathrm{Re}\,G_0^- = \mathbf{I}\nabla_k\mathbf{G}_1\nabla_Y\Pi - \nabla_k\Pi\nabla_Y(\mathbf{IG}_1) - \frac{\mathbf{IG}_1}{a_{33}}\mathrm{div}_Y(a_{33}\nabla_k\Pi),$$
which along with the equality $\mathbf{I}a_{33} = \sqrt{\det\mathbb{G}}$ implies
$$(2\mathbf{G}_1)\,\mathrm{Re}\,G_0^- = \nabla_k\mathbf{G}_1\cdot\nabla_Y\Pi - \nabla_k\Pi\cdot\nabla_Y\mathbf{G}_1 - \mathbf{G}_1\,\mathrm{div}_Y(\nabla_k\Pi) - \mathbf{G}_1\nabla_k\Pi\cdot\nabla_Y\ln\sqrt{\det\mathbb{G}}.$$
Noting that
$$\mathrm{div}_Y(\nabla_k\Pi) + \nabla_k\Pi\cdot\nabla_Y\ln\sqrt{\det\mathbb{G}} = \mathbf{div}(\nabla_k\Pi),$$
and recalling
(G.39)
$$G_0 = -\frac{a_{33}}{J}G_0^- = -\frac{\det\mathbb{G}}{J^2}G_0^-,$$
we obtain
$$\mathrm{Re}\,G_0 = \frac{\det\mathbb{G}}{2J^2\mathbf{G}_1}(\nabla_k\Pi\cdot\nabla_Y\mathbf{G}_1 - \nabla_k\mathbf{G}_1\cdot\nabla_Y\Pi) + \frac{\det\mathbb{G}}{2J^2}\mathbf{div}(\nabla_k\Pi).$$
From this and the identities
$$\nabla_k\Pi = \mathbb{G}^{-1}\nabla_Y\tilde{\eta}, \quad \nabla_k\mathbf{G}_1 = \frac{1}{\mathbf{G}_1}\mathbb{G}^{-1}k, \quad \nabla_Y\mathbf{G}_1 = \frac{1}{2\mathbf{G}_1}\nabla_Y(\mathbb{G}^{-1}k\cdot k)$$

we obtain the desired formula (3.19) for the real part of G_0. Next (G.38) yields

$$(2\mathbf{IG}_1)\,\operatorname{Im} G_0^- = -\mathbf{I}\nabla_k \mathbf{G}_1 \cdot \nabla_Y(\mathbf{IG}_1) - \nabla_k \Pi \cdot \nabla_Y \Pi -$$
$$\frac{1}{a_{33}}\Pi \operatorname{div}_Y(a_{33}\nabla_k \Pi) + \frac{1}{a_{33}}\operatorname{div}_Y\bigl[\Pi\nabla_k(a_{33}\Pi) + J\mathbf{G}_1\nabla_k\mathbf{G}_1\bigr] =$$
$$\frac{J}{a_{33}}\mathbf{G}_1 \operatorname{div}_Y(\nabla_k\mathbf{G}_1) + \mathbf{G}_1\Bigl(\frac{1}{a_{33}}\nabla_Y J - \mathbf{I}\nabla_Y \mathbf{I}\Bigr)\cdot\nabla_k\mathbf{G}_1.$$

Since $J/a_{33} = \mathbf{I}^2$ and

$$\frac{1}{a_{33}}\nabla_Y J - \mathbf{I}\nabla_Y \mathbf{I} = \frac{J^2}{(\det \mathbb{G})^{3/2}}\nabla_Y \sqrt{\det \mathbb{G}} = \mathbf{I}^2 \frac{1}{\sqrt{\det \mathbb{G}}}\nabla_Y \sqrt{\det \mathbb{G}},$$

we have

$$2\operatorname{Im} G_0^- = \mathbf{I}\,\operatorname{\mathbf{div}}(\nabla_k \mathbf{G}_1).$$

Recalling (G.39) we obtain (3.20) and the theorem follows.

Invariant form of $\operatorname{Re} G_0$. In the rest of the chapter we prove the formula (3.24). We start with the calculation of the quadratic form

$$Q(Y, \mathbb{G}\xi) := \tilde{Q}(Y, \xi) = \tilde{Q}_{11}\xi_1^2 + 2\tilde{Q}_{12}\xi_1\xi_2 + \tilde{Q}_{22}\xi_2^2.$$

It follows from (3.21) and the identity $\mathbb{G}\partial_{y_j}\mathbb{G}^{-1}\mathbb{G} = -\partial_{y_j}\mathbb{G}$ that

$$(G.40) \qquad \tilde{Q}(Y, \xi) = \sum_{j=1}^{2}(\xi_j\partial_{y_j}\mathbb{G})\mathbf{q}\cdot\xi - \frac{1}{2}\partial_{y_j}(\mathbb{G}\xi\cdot\xi)q_j - \nabla_Y^2\tilde{\eta}\xi\cdot\xi,$$

where the vector field $\mathbf{q} = \mathbb{G}^{-1}\nabla_Y \tilde{\eta}$. Thus we get

$$\tilde{Q}_{\alpha\beta} = \frac{1}{2}\sum_{j=1}^{2}\Bigl(\partial_{y_\alpha}g_{j\beta} + \partial_{y_\beta}g_{j\alpha} - \partial_{y_j}g_{\alpha\beta}\Bigr)q_j - \partial_{\alpha\beta}^2\tilde{\eta},$$

which along with the equality $g_{\alpha\beta} = \partial_{y_\alpha}\mathbf{r}\cdot\partial_{y_\beta}\mathbf{r}$ yields

$$(G.41) \qquad \tilde{Q}_{\alpha\beta} = \Bigl(\sum_{j=1}^{2}q_j\partial_{y_j}\mathbf{r} - \mathbf{e}_3\Bigr)\cdot\partial_{y_\alpha y_\beta}^2\mathbf{r}.$$

On the other hand, since $\tilde{\eta} = \mathbf{r}\cdot\mathbf{e}_3$, the expression for $\mathbf{q}$ reads

$$\mathbf{q} = \frac{1}{\det \mathbb{G}}\left\{\begin{array}{c}(g_{22}\partial_{y_1}\mathbf{r} - g_{12}\partial_{y_2}\mathbf{r})\cdot\mathbf{e}_3 \\ (-g_{12}\partial_{y_1}\mathbf{r} + gg_{11}\partial_{y_2}\mathbf{r})\cdot\mathbf{e}_3\end{array}\right\}.$$

Now set

$$(G.42) \qquad \mathbf{a} = \partial_{y_1}\mathbf{r}, \quad \mathbf{b} = \partial_{y_2}\mathbf{r}, \quad \mathbf{c} = \mathbf{a}\times\mathbf{b}.$$

Noting that

$$g_{11} = \mathbf{a}\cdot\mathbf{a}, \quad g_{12} = \mathbf{a}\cdot\mathbf{b}, \quad g_{22} = \mathbf{b}\cdot\mathbf{b},$$
$$\mathbf{b}\times\mathbf{c} = (\mathbf{b}\cdot\mathbf{b})\mathbf{a} - (\mathbf{a}\cdot\mathbf{b})\mathbf{b}, \quad \mathbf{a}\times\mathbf{c} = (\mathbf{a}\cdot\mathbf{b})\mathbf{a} - (\mathbf{a}\cdot\mathbf{a})\mathbf{b},$$

we obtain

$$(\det \mathbb{G})q_1 = (\mathbf{b}\times\mathbf{c})\cdot\mathbf{e}_3, \quad (\det \mathbb{G})q_2 = -(\mathbf{a}\times\mathbf{c})\cdot\mathbf{e}_3.$$

From this and the identity
$$[(\mathbf{b} \times \mathbf{c}) \cdot \mathbf{e}_3]\mathbf{a} - [(\mathbf{a} \times \mathbf{c}) \cdot \mathbf{e}_3]\mathbf{b} - |\mathbf{c}|^2 \mathbf{e}_3 = -(\mathbf{c} \cdot \mathbf{e}_3)\mathbf{c},$$
which holds true for all $\mathbf{a}, \mathbf{b} \in \mathbb{R}^3$ and $\mathbf{c} = \mathbf{a} \times \mathbf{b}$, we conclude that
$$(\det \mathbb{G})(q_1 \partial_{y_1}\mathbf{r} + q_2 \partial_{y_2}\mathbf{r}) - |\mathbf{c}|^2 \mathbf{e}_3 = -(\mathbf{c} \cdot \mathbf{e}_3)\mathbf{c}.$$
Noting that $|\mathbf{c}|^2 = \det \mathbb{G}$ and $\mathbf{c} \cdot \mathbf{e}_3 = J$ we arrive at
$$\sum_{j=1}^{2} q_j \partial_{y_j}\mathbf{r} - \mathbf{e}_3 = -\frac{J}{\sqrt{\det \mathbb{G}}}\mathbf{n},$$
where $\mathbf{n} = \mathbf{c}/|\mathbf{c}|$ is the unit normal vector to Σ. Substituting this identity into (G.41) gives $\tilde{Q}_{\alpha\beta} = -(\mathbf{n} \cdot \partial_{y_\alpha y_\beta}\mathbf{r})J/\sqrt{\det \mathbb{G}}$ which leads to
$$\tilde{Q}(Y,\xi) = -\frac{J}{\sqrt{\det \mathbb{G}}}(L\xi_1^2 + 2M\xi_1\xi_2 + N\xi_2^2).$$
Since
$$\mathbf{G}_1^2(Y, \mathbb{G}\xi) = \mathbb{G}\xi \cdot \xi := E\xi_1^2 + 2F\xi_1\xi_2 + G\xi_2^2,$$
we finally obtain
(G.43)
$$\frac{\det \mathbb{G}}{2J^2}\left\{\frac{1}{\mathbf{G}_1^2(Y, \mathbb{G}\xi)}Q(Y, \mathbb{G}\xi)\right\} = -\frac{\sqrt{\det \mathbb{G}}}{J}\frac{L\xi_1^2 + 2M\xi_1\xi_2 + N\xi_2^2}{2(E\xi_1^2 + 2F\xi_1\xi_2 + G\xi_2^2)}.$$

Our next task is to express $\mathbf{div}\,\mathbf{q}$ via the geometric characteristics of Σ. First we do this in the standard coordinates $Y = X$ with $\tilde{\eta} = \eta$. In this case
$$\mathbb{G}^{-1} = \frac{1}{1+|\nabla\eta|^2}\begin{pmatrix} 1+\partial_{y_2}\eta^2, & -\partial_{y_1}\eta\,\partial_{y_2}\eta \\ -\partial_{y_1}\eta\,\partial_{y_2}\eta, & 1+\partial_{y_1}\eta^2 \end{pmatrix}, \quad \det \mathbb{G} = 1+|\nabla\eta|^2, \quad J = 1,$$
and $\mathbf{q} = (1+|\nabla\eta|^2)^{-1}\nabla\eta$, which leads to the formula
(G.44) $$\frac{\det \mathbb{G}}{2J^2}\mathbf{div}\,\mathbf{q} = \frac{\sqrt{\det \mathbb{G}}}{J}\frac{1}{2}\mathrm{div}\left(\frac{\nabla\eta}{\sqrt{1+|\nabla\eta|^2}}\right) = \frac{\sqrt{\det \mathbb{G}}}{J}\frac{\mathfrak{k}_1+\mathfrak{k}_2}{2},$$
where $\mathfrak{k}_i$ are the principal curvatures of Σ at the point $\mathbf{r}$. Next note that $\nabla\eta$ is a covariant vector field on Σ, hence $\mathbb{G}^{-1}\nabla\eta$ is a vector field on Σ. Since $\mathbf{div}$ is an invariant operator on the space of vector fields on Σ, the left side of (G.44) does not depend on the choice of coordinates, hence
(G.45) $$\frac{\det \mathbb{G}}{2J^2}\mathbf{div}\,\mathbf{q} = \frac{\sqrt{\det \mathbb{G}}}{J}\frac{\mathfrak{k}_1+\mathfrak{k}_2}{2} = \frac{\sqrt{\det \mathbb{G}}}{J}\frac{LG - 2MF + NE}{2(EG - F^2)}.$$
Combining (G.43) and (G.45) gives the desired identity (3.24). If we define by $\mathfrak{n}(\xi)$ the normal curvature of Σ in the direction ξ at a point $\mathbf{r}$, then (3.24) becomes
$$\frac{J}{\sqrt{\det \mathbb{G}}}\,\mathrm{Re}\,G_0(Y, \mathbb{G}\xi) = \frac{1}{2}(\mathfrak{k}_1+\mathfrak{k}_2-\mathfrak{n}(\xi)),$$
which leads to

COROLLARY G.3. *Assume that the manifold Σ has a parametric representation $\mathbf{r}(Y) = (\mathbb{T}Y + \tilde{\mathcal{V}}(Y), \tilde{\eta}(Y))$, $Y \in \mathbb{R}^2$ so that $\tilde{\mathcal{V}}$ and $\tilde{\eta}$, which are not defined yet, satisfy all hypotheses of Theorem 3.5. Assume also that the parametric form*

$$\mathfrak{G}u := \frac{J}{\sqrt{\det \mathbb{G}}} \mathcal{G}_\eta \check{u} \circ (\mathbb{T} + \mathcal{V}), \quad \check{u}(X) = u(Y(X)),$$

of the normal derivative operator is given for any bi-periodic smooth function $u(Y)$. Then the manifold Σ is defined by the operator $\mathfrak{G}$ up to a translation and a rotation of the embedding space.

PROOF. Note that for all $k \in \mathbb{Z}^2$,

$$\lim_{n \to \infty} \frac{1}{n} e^{-ink \cdot Y} \mathfrak{G} e^{ink \cdot Y} = \frac{J}{\sqrt{\det \mathbb{G}}} G_1(Y, k) = \sqrt{\mathbb{G}^{-1} k \cdot k},$$

$$\lim_{n \to \infty} \operatorname{Re} \left\{ e^{-ink \cdot Y} \mathfrak{G} e^{ink \cdot Y} - \frac{J}{\sqrt{\det \mathbb{G}}} G_1(Y, k) \right\} = \frac{J}{\sqrt{\det \mathbb{G}}} \operatorname{Re} G_0(Y, k).$$

Since G_0 is a homogeneous function of k, it follows from this that the right hand sides of these equalities are defined by the operator $\mathfrak{G}$ for all $k \in \mathbb{R}^2$ and, in particular, for $k = \mathbb{G}\xi$ with an arbitrary $\xi \in \mathbb{R}^2$. Hence the first fundamental form $\mathbb{G}\xi \cdot \xi$ and the difference $\mathfrak{k}_1 + \mathfrak{k}_2 - \mathfrak{n}(\xi)$ are completely defined by $\mathfrak{G}$ for all directions ξ at each point of Σ. Hence the principle curvatures of Σ are also defined by the operator $\mathfrak{G}$. It remains to note that, by the Bonnet Theorem, the first fundamental form and the principal curvatures define Σ up to a translation and a rotation of the embedding space. $\square$

APPENDIX H

Proof of Lemma 5.8

To be able to compute all terms in (5.95), let us rewrite the system (1.6), (1.7) formally as a scalar equation for ψ and express the orders ε and ε^2 of the differential computed at $\psi_\varepsilon^{(N)}$, $N \geq 3$. Then the operator $\mathfrak{L} + \mathfrak{H}$ is up to order ε^2 closely linked with the new form of this operator after applying the diffeomorphism computed at Lemma 3.8.

Indeed we can formally solve (1.7) with respect to η in powers of ψ as

$$\text{(H.1)} \qquad \eta = -\frac{1}{\mu}\partial_{x_1}\psi - \frac{1}{2\mu}(\nabla\psi)^2 + \frac{1}{2\mu^3}(\partial_{x_1}^2\psi)^2 + O(\|\psi\|^3),$$

and replace η by this expression in (1.6). We then obtain a new scalar *formal equation* for ψ, under the form $\mathcal{E}(\psi, \mu) = 0$, where

$$\mathcal{E}(\psi,\mu) = \mathfrak{L}_0\psi + (\mu-\mu_c)\mathfrak{L}_1\psi + \mathcal{E}_2(\psi,\psi) + \mathcal{E}_3(\psi,\psi,\psi) + O(|\mu-\mu_c|^2\|\psi\| + \|\psi\|^4),$$

where

$$\mathfrak{L}_0\psi = \frac{1}{\mu_c}\partial_{x_1}^2\psi + \mathcal{G}^{(0)}\psi$$

$$\mathfrak{L}_1\psi = -\frac{1}{\mu_c^2}\partial_{x_1}^2\psi,$$

and $\mathcal{E}_2$ and $\mathcal{E}_3$ represent quadratic and cubic terms in ψ, respectively. Let us write the formal solution found at Theorem 2.3 for $\varepsilon_1 = \varepsilon_2 = \varepsilon/2$, under the form ($N \geq 3$)

$$\psi_\varepsilon^{(N)} = \varepsilon\psi_1 + \varepsilon^2\psi_2 + O(\varepsilon^3),$$
$$\mu = \mu_c + \varepsilon^2\mu_1 + O(\varepsilon^3),$$

where

$$\psi_1 = \sin x_1 \cos \tau x_2,$$

then we have the identities

$$\mathfrak{L}_0\psi_1 = 0,$$
$$\text{(H.2)} \qquad \mathfrak{L}_0\psi_2 + \mathcal{E}_2(\psi_1,\psi_1) = 0,$$
$$\mathfrak{L}_0\psi_3 + \mu_1\mathfrak{L}_1\psi_1 + 2\mathcal{E}_2(\psi_1,\psi_2) + \mathcal{E}_3(\psi_1,\psi_1,\psi_1) = 0.$$

Now, we may observe that the operator (3.7) we want to invert acts on $\delta\phi = \delta\psi - \mathfrak{b}\delta\eta$. Since $\mathfrak{b}$ is $O(\varepsilon)$ and $\delta\eta$ may be expressed formally linearly in terms of $\delta\psi$ in differentiating formally (H.1), we have formally

$$\text{(H.3)} \qquad \partial_\psi \mathcal{E}(\psi,\mu)(1 + \mathcal{H}(\psi,\mu)) = \kappa \mathfrak{L}(\psi,\mu) - \mathcal{R}(\mathcal{E},\mu)$$

where $\mathfrak{L}(\psi,\mu)$ is the linear operator we want to invert, κ is the function we introduced at Theorem 3.4, and the operator $\mathcal{H}(\psi,\mu)$ is such that formally
$$\delta\psi = (1+\mathcal{H}(\psi,\eta))\delta\phi,$$
and $\mathcal{R}(0,\mu) = 0$. Since we set
$$\psi = \psi_\varepsilon^{(N)} + O(\varepsilon^N), \ N \geq 3$$
$$\mu = \mu_0 + \varepsilon^2 \mu_1(\tau_0) + O(\varepsilon^3), \ \mu_0 = \mu_c(\tau_0),$$
we have
$$\mathcal{E}(\psi_\varepsilon^{(2)}, \mu_0 + \varepsilon^2 \mu_1) = O(\varepsilon^3), \quad \mathcal{R} = O(\varepsilon^3),$$
hence
$$\partial_\psi \mathcal{E}(\psi,\mu) = \mathfrak{L}_0 + \varepsilon^2 \mu_1 \mathfrak{L}_1 + 2\mathcal{E}_2(\psi_\varepsilon^{(2)},\cdot) + 3\mathcal{E}_3(\psi_\varepsilon^{(2)},\psi_\varepsilon^{(2)},\cdot) + O(\varepsilon^3).$$

Making now the change of coordinates computed at Lemma 3.8, the new expressions of operators $\mathfrak{L}_0$, $\mathfrak{L}_1$, $\mathcal{E}_2(\psi_\varepsilon^{(2)},\cdot)$, $\mathcal{E}_3(\psi_1,\psi_1,\cdot)$ take the following form
$$\text{new}\mathfrak{L}_0 = \mathfrak{L}_0 + \varepsilon \mathfrak{L}_0^{(1)} + \varepsilon^2 \mathfrak{L}_0^{(2)} + O(\varepsilon^3),$$
$$\text{new}\mathfrak{L}_1 = \mathfrak{L}_1^{(0)} + \varepsilon \mathfrak{L}_1^{(1)} + O(\varepsilon^2),$$
$$\text{new}\mathcal{E}_2(\psi_1,\cdot) = \mathcal{E}_2^{(0)}(\psi^{(0)},\cdot) + \varepsilon \mathcal{E}_2^{(1)}(\psi^{(0)},\cdot) + O(\varepsilon^2),$$
$$\text{new}\mathcal{E}_3(\psi_1,\psi_1,\cdot) = \mathcal{E}_3^{(0)}(\psi^{(0)},\psi^{(0)},\cdot) + O(\varepsilon),$$
and the functions ψ_1, ψ_2 are transformed into
$$\text{new } \psi_1 = \psi^{(0)} + \varepsilon \psi_1^{(1)} + \varepsilon^2 \psi_1^{(2)} + O(\varepsilon^3)$$
$$\text{new}\psi_2 = \psi_2^{(0)} + \varepsilon \psi_2^{(1)} + O(\varepsilon^2).$$
Moreover, we have thanks to Lemma 3.8
$$\kappa(Y) = 1 + \varepsilon \kappa_1(Y) + \varepsilon^2 \kappa_2(Y) + O(\varepsilon^3),$$
$$\text{new}\mathcal{H}(\psi,\mu) = \varepsilon \mathcal{H}_1 + \varepsilon^2 \mathcal{H}_2 + O(\varepsilon^3).$$
Hence, identities (H.2) lead to

(H.4)
$$\mathfrak{L}_0 \psi^{(0)} = 0,$$
$$\mathfrak{L}_0^{(1)} \psi^{(0)} + \mathfrak{L}_0 \psi_1^{(1)} = 0,$$
$$\mathfrak{L}_0^{(2)} \psi^{(0)} + \mathfrak{L}_0^{(1)} \psi_1^{(1)} + \mathfrak{L}_0 \psi_1^{(2)} = 0,$$

(H.5)
$$\mathfrak{L}_0 \psi_2^{(0)} + \mathcal{E}_2^{(0)}(\psi^{(0)}, \psi^{(0)}) = 0,$$
$$\mathfrak{L}_0^{(1)} \psi_2^{(0)} + \mathfrak{L}_0 \psi_2^{(1)} + 2\mathcal{E}_2^{(0)}(\psi^{(0)}, \psi_1^{(1)}) + \mathcal{E}_2^{(1)}(\psi^{(0)}, \psi^{(0)}) = 0,$$

(H.6) $\quad \mathfrak{L}_0 \psi_3^{(0)} + \mu_1 \mathfrak{L}_1^{(0)} \psi^{(0)} + 2\mathcal{E}_2^{(0)}(\psi^{(0)}, \psi_2^{(0)}) + \mathcal{E}_3^{(0)}(\psi^{(0)}, \psi^{(0)}, \psi^{(0)}) = 0.$

We deduce from these formulae and from (H.3), that the linear operator obtained after the change of coordinates satisfies
$$\mathfrak{L} + \mathfrak{H} = \mathfrak{L}_0 + \varepsilon \mathfrak{H}^{(1)} + \varepsilon^2 \mathfrak{H}^{(2)} + O(\varepsilon^3)$$

with
(H.7) $$\kappa_1 \mathfrak{L}_0 + \mathfrak{H}^{(1)} = \mathfrak{L}_0^{(1)} + 2\mathcal{E}_2^{(0)}(\psi^{(0)}, \cdot) + \mathfrak{L}_0 \mathcal{H}_1$$

$$\kappa_2 \mathfrak{L}_0 + \kappa_1 \mathfrak{H}^{(1)} + \mathfrak{H}^{(2)} = \mathfrak{L}_0^{(2)} + \mu_1 \mathfrak{L}_1^{(0)} + 2\mathcal{E}_2^{(1)}(\psi^{(0)}, \cdot) + 2\mathcal{E}_2^{(0)}(\psi_2^{(0)}, \cdot) + 2\mathcal{E}_2^{(0)}(\psi_1^{(1)}, \cdot) +$$
$$+ 3\mathcal{E}_3^{(0)}(\psi^{(0)}, \psi^{(0)}, \cdot) + \left\{\mathfrak{L}_0^{(1)} + 2\mathcal{E}_2^{(0)}(\psi^{(0)}, \cdot)\right\} \mathcal{H}_1 + \mathfrak{L}_0 \mathcal{H}_2.$$

We now compute the terms under the integral in (5.95). First we observe (thanks to (H.4), (H.5))

$$\begin{aligned}\mathfrak{L}_0^{-1} \mathfrak{H}^{(1)} \psi^{(0)} &= \mathfrak{L}_0^{-1} \{\mathfrak{L}_0^{(1)} \psi^{(0)} + 2\mathcal{E}_2^{(0)}(\psi^{(0)}, \psi^{(0)})\} + \mathcal{H}_1 \psi^{(0)} \\ &= \mathfrak{L}_0^{-1} \{-\mathfrak{L}_0 \psi_1^{(1)} - 2\mathfrak{L}_0 \psi_2^{(0)}\} + \mathcal{H}_1 \psi^{(0)} \\ &= -(\psi_1^{(1)} + 2\psi_2^{(0)}) + \mathcal{H}_1 \psi^{(0)},\end{aligned}$$

hence

$$\begin{aligned}-\mathfrak{H}^{(1)} \mathfrak{L}_0^{-1} \mathfrak{H}^{(1)} \psi^{(0)} &= \mathfrak{L}_0^{(1)}(\psi_1^{(1)} + 2\psi_2^{(0)}) + 2\mathcal{E}_2^{(0)}(\psi^{(0)}, \psi_1^{(1)}) + \\ &+ 4\mathcal{E}_2^{(0)}(\psi^{(0)}, \psi_2^{(0)}) + \mathfrak{L}_0 \mathcal{H}_1(\psi_1^{(1)} + 2\psi_2^{(0)}) + \\ &- \kappa_1 \mathfrak{L}_0(\psi_1^{(1)} + 2\psi_2^{(0)}) - \mathfrak{H}^{(1)} \mathcal{H}_1 \psi^{(0)}.\end{aligned}$$

Moreover
$$\begin{aligned}\mathfrak{H}^{(2)} \psi^{(0)} &= \mathfrak{L}_0^{(2)} \psi^{(0)} + \mu_1 \mathfrak{L}_1^{(0)} \psi^{(0)} + 2\mathcal{E}_2^{(1)}(\psi^{(0)}, \psi^{(0)}) + 2\mathcal{E}_2^{(0)}(\psi_2^{(0)}, \psi^{(0)}) + \\ &+ 2\mathcal{E}_2^{(0)}(\psi_1^{(1)}, \psi^{(0)}) + 3\mathcal{E}_3^{(0)}(\psi^{(0)}, \psi^{(0)}, \psi^{(0)}) - \kappa_1 \mathfrak{H}^{(1)} \psi^{(0)} + \\ &+ \left\{\mathfrak{L}_0^{(1)} + 2\mathcal{E}_2^{(0)}(\psi^{(0)}, \cdot)\right\} \mathcal{H}_1 \psi^{(0)} + \mathfrak{L}_0 \mathcal{H}_2 \psi^{(0)}\end{aligned}$$

and in using again (H.4), (H.5), (H.6), and (H.7) we obtain

$$\begin{aligned}\mathfrak{H}^{(2)} \psi^{(0)} - \mathfrak{H}^{(1)} \mathfrak{L}_0^{-1} \mathfrak{H}^{(1)} \psi^{(0)} &= -2\mu_1 \mathfrak{L}_1^{(0)} \psi^{(0)} - \mathfrak{L}_0(3\psi_3^{(0)} + 2\psi_2^{(1)} + \psi_1^{(2)}) + \\ &+ \mathfrak{L}_0 \{\mathcal{H}_1(\psi_1^{(1)} + 2\psi_2^{(0)}) + \mathcal{H}_2 \psi^{(0)} - \mathcal{H}_1^2 \psi^{(0)}\}.\end{aligned}$$

Hence (5.95) leads to

$$@ = -2\mu_1 \int_{\mathbb{T}^2} (\mathfrak{L}_1^{(0)} \psi^{(0)}) \psi^{(0)} dY.$$

Since
$$\mathfrak{L}_1^{(0)} = -\frac{1}{\mu_0^2} \partial_{y_1}^2$$

we finally obtain the result of Lemma 5.8.

APPENDIX I

Fluid particles dynamics

The kinematic and dynamic boundary conditions (1.2, 1.3) give two equations for two unknowns η and ψ. Assume for the moment that we know η i.e. the free surface Σ. The question is *can we restore ψ without solving PDE equations?* The answer is yes: it suffices to solve a problem of moving a heavy single mass point along the free surface, or equivalently to find the corresponding geodesic flow on the surface with an appropriate metric.

Let us begin with the consideration of the motion of a single mass point along the surface $\Sigma = \{x_3 = \eta(X)\}$. Assuming that gravity μ acts in the $-e_3 = (0,0,-1)$ direction we can write the governing equations in the form

$$\ddot{x} + \mu e_3 = \lambda \mathbf{n}, \quad x_3 = \eta(X),$$

where λ is the Lagrange multiplier and $\mathbf{n}$ is a normal vector to Σ. Choosing components of X as generalized coordinates we rewrite equivalently these equation in the Lagrange form with the Lagrangian

$$\mathbf{L}(X, \dot{X}) = \mathbf{T}(X, \dot{X}) - \mathbf{U}(X) = \frac{1}{2}\mathbb{G}(X)\dot{X} \cdot \dot{X} - \mu\eta(X),$$

where $\mathbb{G}dX \cdot dX$ is the first fundamental form of the free surface Σ with

$$\mathbb{G}(X) = I + \nabla_X \eta(X) \otimes \nabla_X \eta(X).$$

More precisely, we have

$$\frac{d}{dt}\partial_{\dot{X}}\mathbf{L}(X, \dot{X}) - \partial_X \mathbf{L}(X, \dot{X}) = 0.$$

If we define the moments and Hamiltonian by

$$y = \mathbb{G}\dot{X}, \quad \mathbf{H}(X, y) = \frac{1}{2}\mathbb{G}^{-1}y \cdot y + \mu\eta(X),$$

then the governing equations can be rewritten in the Hamilton form

(I.1) $$\dot{X} = \partial_y \mathbf{H}(X, y), \quad \dot{y} = -\partial_X \mathbf{H}(X, y).$$

Next, suppose that a C^1-generating function $S(X)$ satisfies the Hamilton-Jacobi equation

(I.2) $$\mathbf{H}(X, \nabla_X S(X)) = h = \text{const},$$

and the periodicity conditions

(I.3) $$S(X + 2\pi e_1) - 2\pi = S(X + \frac{2\pi}{\tau}e_2) - \frac{2\pi}{\tau} = S(X).$$

Suppose also that $X(t)$ is a solution of the equations

(I.4) $$\dot{X} = \mathbb{G}^{-1}(X)\nabla_X S(X),$$

then, it is known that $(X(t), y(t))$, with $y(t) = \nabla_X S(X(t))$, is a solution of the Hamiltonian system (I.1), and the surface $y = \nabla_X S(X)$ is an invariant manifold of (I.1), the flow being defined by (I.4).

Finally note that due to the periodicity conditions, the mapping $X \to (X, \nabla_X S(X))$ defines an embedding of the torus $\mathbb{R}^2/\Gamma$ into $\mathbb{R}^2/\Gamma \times \mathbb{R}^2$. Therefore, $\{y = \nabla_X S(X)\}$ is an invariant torus of (I.1) lying on the energy surface $H = h$. In coordinates X the Hamiltonian flow on the torus is given by (I.4)

Let us turn to the diamond wave problem. Set

$$\varphi^*(x) = \mathbf{u}_0 \cdot X + \varphi(x), \text{ and } \psi^*(X) = \varphi^*(x_1, x_2, \eta(X)),$$

and recall $|\mathbf{u}_0| = 1$. In these notations kinematic condition (1.2) and dynamic condition (1.3) can be rewritten in the equivalent form

(I.5) $$\nabla_X \varphi^*(x_1, x_2, x_3) = \mathbb{G}(X)^{-1}\nabla_X \psi^*(X) \text{ for } x_3 = \eta(X),$$

(I.6) $$\frac{1}{2}\mathbb{G}^{-1}\nabla_X \psi^* \cdot \nabla_X \psi^* + \mu\eta = \frac{1}{2}.$$

By construction the equation $\dot{x} = \nabla\varphi^*(x)$ determines the trajectories of liquid particles. From (I.5), such a particle moving along the free surface satisfies

(I.7) $$\dot{X} = G^{-1}(X)\nabla_X \psi^*(X).$$

On the other hand, equation (I.6) reads

$$\mathbf{H}(X, \nabla_X \psi^*) = 1/2.$$

Hence $S(X) = \psi^*(X)$ is a generating function for the Hamiltonian system (I.1).

From this we conclude that trajectories of liquid particles $X(t)$, which are defined by $\dot{X} = \mathbb{G}^{-1}\nabla_X \psi^*(X)$, along with $y(t) = \nabla_X S(X(t))$ serve as solutions of (I.1) and belongs to the invariant torus $\{y = \nabla_X S(X)\}$. In other words, they coincide with projections $(X, y) \to X$ of solutions to (I.1) belonging to the invariant torus $\{y = \nabla\psi^*(X)\} \subset \{\mathbf{H} = 1/2\}$. Moreover, since by (3.1) we have $V = \mathbb{G}^{-1}\nabla_X\psi^*$, they also coincide with the integral curves of the vector field V.

Finally note that, by the Maupertuis principle, the projections $(X, y) \to X$ of solutions of (I.1) belonging to the energy surface $\mathbf{H} = 1/2$, coincide with the geodesics on the manifold Σ endowed with the Jacobi metric

(I.8) $$ds^2 = (1/2 - \mu\eta(X))\mathbb{G}(X)\,dX \cdot dX \equiv 2(1/2 - \mathbf{U})\mathbf{T}(X, dX).$$

Hence the integral curves of the vector field V form the geodesic flow associated with the metric (I.8).

COROLLARY I.1. *Suppose that η is an arbitrary bi-periodic smooth function so that the hamiltonian system (I.1) has an invariant torus $\{y = \nabla_X S(X)\}$ with a smooth generating function $S(X)$ satisfying the periodicity conditions (I.3). Then the solution φ^* of the Cauchy problem*

$$\varphi^*(x) = S(X), \quad \partial_n \varphi^*(x) = 0 \text{ for } x_3 = \eta(X),$$

for the Laplace equation, satisfies kinematic and dynamic conditions (1.2),(1.3). (The local existence of such a solution follows from the Cauchy-Kowalewski theorem, and the existence and boundedness in the lower half plane is true only for the "good" choice of generating function.)

Bibliography

[ABB] X. Antoine, H. Barucq, A.Bendali. *Bayliss-Turkel- like Radiation Conditions on Surfaces of Arbitrary Shape.* J. Math. Anal. Appl. **229** (1999), 184-211.

[A] V.I. Arnold. *Proof of a theorem of A.N. Kolmogorov on the invariance of quasi-periodic motions under small perturbations of the Hamiltonian.* Russ. Math. Surv. **18** (1963), 9-36.

[B1] J. Bourgain. *Construction of periodic solutions of nonlinear wave equations in higher dimension.* Geom. Funct. Anal. **5** (1995), 629-639.

[B2] J. Bourgain. *Quasi-periodic solutions of Hamiltonian perturbations of 2D linear Schrödinger equations.* Ann. of Math. **148** (1998), 363-439

[BDM] T.Bridges, F.Dias, D.Menasce. *Steady three-dimensional water-wave patterns on a finite-depth fluid.* J.Fluid Mech. **436** (2001), 145-175.

[Ca] J.W.S. Cassels, *An Introduction to Diophantine approximations,* Cambridge University Press, 1957.

[Cr] W. Craig, *Problèmes de petits diviseurs dans les équations aux dérivées partielle,* Panoramas et Synthèses **9**, Société Mathématique de France, Paris, 2000.

[CrN1] W.Craig, D.Nicholls. *Traveling gravity water waves in two and three dimensions.* EJMB/Fluids **21** (2002), 615-641.

[CrN2] W.Craig, D.Nicholls. *Travelling two and three-dimensional capillary gravity water waves.* SIAM J. Math. Anal. **32** (2000), 323-359.

[CrSS] W.Craig, U.Schanz, C.Sulem. *The modulational regime of three-dimensional water waves and the Davey-Stewartson system.* Ann. Inst. Henri Poincaré **14**, 5 (1997), 615-667.

[CrWa] W. Craig, E. Wayne. *Newton's method and periodic solutions of nonlinear wave equation.* Comm. Pure Applied Math. **XLVI** (1993), 1409-1501.

[D] K. Deimling. *Nonlinear Functional Analysis,* Springer-Verlag, Heidelberg, 1985

[DiIo] F.Dias, G.Iooss, *Water waves as a spatial dynamical system.* Handbook of Mathematical Fluid Dynamics, chap 10, p.443 -499. S.Friedlander, D.Serre Eds., Elsevier, 2003.

[DiK] F.Dias, C.Kharif, *Nonlinear gravity and capillary-gravity waves.* Annu. Rev. Fluid Mech. **31** (1999), 301-346.

[FS] J. Fröhlich, T. Spencer. *Absence of diffusion in the Anderson tight binding model for large disorder or low energy.* Comm. Math. Phys. **88** (1983), 151-184.

[Fu] R.Fuchs. *On the theory of short-crested oscillatory waves.* U.S. Natl. Bur. Stand. Circ. **521** (1952), 187-200.

[G] M.D.Groves.*An existence theory for three-dimensional periodic travelling gravity-capillary water waves with bounded transverse profiles.* Physica D **152-153** (2001), 395-415.

[GH] M.D.Groves, M.Haragus.*A bifurcation theory for three-dimensional oblique travelling gravity-capillary water waves.* J.Nonlinear Sci. **13** (2003), 397-447.

[GM] M.Groves, A.Mielke. *A spatial dynamics approach to three-dimensional gravity-capillary steady water waves.* Proc. Roy. Soc. Edin. A **131** (2001), 83-136.

[HHS] J.Hammack, D.Henderson, H.Segur. *Progressive waves with persistent, two-dimensional surface patterns in deep water.* J.Fluid Mech. **532** (2005), 1-52.

[HK] M.Haragus-Courcelle, K.Kirchgässner. *Three-dimensional steady capillary-gravity waves,* Ergodic theory, Analysis and efficient simulation of dynamical systems (Ed. B.Fiedler), Berlin, Springer-Verlag 2001, pp.363-397,

[Ho] L. Hörmander. *Pseudo-differential operators and non-elliptic boundary problems.* Ann. of Math. **83** (1966), 129-209.

[I] G.Iooss. *Capillary and Capillary-Gravity periodic travelling waves for two superposed fluid layers, one being of infinite depth* . J. Math. Fluid Mech. **1**(1999), 24-61.

[IPT] G.Iooss, P.Plotnikov, J.Toland. *Standing waves on an infinitely deep perfect fluid under gravity.* Arch. Rat. Mech. Anal. **177** (2005), 367-478.

[K] K.Kirchgässner. *Wave solutions of reversible systems and applications.* J.Diff. Eqns. **45** (1982), 113-127.

[KN] J.J. Kohn, L. Nirenberg. *An algebra of pseudodifferential operators.* Comm. Pure. Apll. Math. **18** (1965), 269-305.

[La] D.Lannes. *Well-posedness of the water-waves equations.* J.Amer. Math. Soc. **18** (2005), 605-654.

[Le] T.Levi-Civita. *Détermination rigoureuse des ondes permanentes d'ampleur finie.* Math. Annalen **93** (1925), 264-314.

[M] J. Moser, *Minimal foliation on a torus,* Topics in calculus of variations (Montecatini Terme), Lecture Notes in Math, vol.1365, 1989, pp. 62-99.

[N] A.I.Nekrasov. *On waves of permanent type.* Izv. Ivanovo-Voznesensk. Politekhn. Inst. **3** (1921), 52-65.

[P] P.I. Plotnikov. *Solvability of the problem of spatial gravitational waves on the surface of an ideal fluid.* Dokl. Akad. Nauk SSSR. **251**(1980), 170-171.

[OM] L.V. Ovsiannykov, N.I. Makarenko, V.I. Nalimov, V. Yu. Liapidevskii, P.I. Plotnikov, I.V. Sturova, V.I. Bukreev, V.A. Vladimirov, *Nonlinear problems in the theory of surface and internal waves* (Russian), Nauka, Novosibirsk, 1985.

[Pe] B.E. Petersen, *An introduction to the Fourier transform and pseudodifferential operators,* Pitman Advanced Publishing programm, Boston, London, Melbourne, 1983.

[PY] P.I. Plotnikov, L.N. Yungerman. *Periodic solutions of a weakly linear wave equation with an irrotational ratio of the period to the interval length.* Diff.Equations **24** (1988), 1059-1065.

[PT] P.Plotnikov, J.Toland. *Nash-Moser theory for standing waves.* Arch. Rat. Mech. Anal. **159** (2001), 1-83.

[ReSh] J.Reeder, M.Shinbrot. *Three-dimensional, nonlinear wave interaction in water of constant depth.* Nonlinear Anal., T.M.A. **5** (1981), 3, 303-323.

[RoSc] A.Roberts, L.Schwartz. *The calculation of nonlinear short-crested gravity waves.* Phys. Fluids. **26** (1983), 2388-2392.

[Si] C. L. Siegel. *Vorlesungen Über Himmelsmechanik* (German, Russian), Inostarannaya Literatura, Moskow, 1959.

[Sr] L.Sretenskii. *Spatial problem of determination of steady waves of finite amplitude* (Russian). Dokl. Akad. Nauk SSSR (N.S.) **89** (1953), 25-28.

[Sto] G.G.Stokes. *On the theory of oscillatory waves.* Trans. Camb. Phil. Soc. **8** (1847), 441-455.

[Str] D. Struik. *Détermination rigoureuse des ondes irrotationnelles périodiques dans un canal à profondeur finie.* Math. Ann. 95 (1926), 595-634.

[T] M.E. Taylor. *Pseudodifferential operators,* Princeton, New Jersey, 1981.

[W] H.Weil. *Uber die gleichverteilung der zahlen rood eins.* Math. Ann. **77** (1916), 313-352.

[Z] V.E.Zakharov. *Stability of periodic waves of finite amplitude on the surface of a deep fluid.* Zh. Prikl. Mekh. Tekh. Fiz. **9** (1968), 86-94, J.Appl. Mech. Tech. Phys. **9** (1968), 190-194.

Editorial Information

To be published in the *Memoirs*, a paper must be correct, new, nontrivial, and significant. Further, it must be well written and of interest to a substantial number of mathematicians. Piecemeal results, such as an inconclusive step toward an unproved major theorem or a minor variation on a known result, are in general not acceptable for publication.

Papers appearing in *Memoirs* are generally at least 80 and not more than 200 published pages in length. Papers less than 80 or more than 200 published pages require the approval of the Managing Editor of the Transactions/Memoirs Editorial Board.

As of March 31, 2009, the backlog for this journal was approximately 12 volumes. This estimate is the result of dividing the number of manuscripts for this journal in the Providence office that have not yet gone to the printer on the above date by the average number of monographs per volume over the previous twelve months, reduced by the number of volumes published in four months (the time necessary for preparing a volume for the printer). (There are 6 volumes per year, each usually containing at least 4 numbers.)

A Consent to Publish and Copyright Agreement is required before a paper will be published in the *Memoirs*. After a paper is accepted for publication, the Providence office will send a Consent to Publish and Copyright Agreement to all authors of the paper. By submitting a paper to the *Memoirs*, authors certify that the results have not been submitted to nor are they under consideration for publication by another journal, conference proceedings, or similar publication.

Information for Authors

Memoirs are printed from camera copy fully prepared by the author. This means that the finished book will look exactly like the copy submitted.

Initial submission. The AMS uses Centralized Manuscript Processing for initial submissions. Authors should submit a PDF file using the Initial Manuscript Submission form found at www.ams.org/peer-review-submission, or send one copy of the manuscript to the following address: Centralized Manuscript Processing, MEMOIRS OF THE AMS, 201 Charles Street, Providence, RI 02904-2294 USA. If a paper copy is being forwarded to the AMS, indicate that it is for it Memoirs and include the name of the corresponding author, contact information such as email address or mailing address, and the name of an appropriate Editor to review the paper (see the list of Editors below).

The paper must contain a *descriptive title* and an *abstract* that summarizes the article in language suitable for workers in the general field (algebra, analysis, etc.). The *descriptive title* should be short, but informative; useless or vague phrases such as "some remarks about" or "concerning" should be avoided. The *abstract* should be at least one complete sentence, and at most 300 words. Included with the footnotes to the paper should be the 2000 *Mathematics Subject Classification* representing the primary and secondary subjects of the article. The classifications are accessible from www.ams.org/msc/. The list of classifications is also available in print starting with the 1999 annual index of *Mathematical Reviews*. The Mathematics Subject Classification footnote may be followed by a list of *key words and phrases* describing the subject matter of the article and taken from it. Journal abbreviations used in bibliographies are listed in the latest *Mathematical Reviews* annual index. The series abbreviations are also accessible from www.ams.org/msnhtml/serials.pdf. To help in preparing and verifying references, the AMS offers MR Lookup, a Reference Tool for Linking, at www.ams.org/mrlookup/.

Electronically prepared manuscripts. The AMS encourages electronically prepared manuscripts, with a strong preference for $\mathcal{A}_{\mathcal{M}}\mathcal{S}$-LaTeX. To this end, the Society has prepared $\mathcal{A}_{\mathcal{M}}\mathcal{S}$-LaTeX author packages for each AMS publication. Author packages include instructions for preparing electronic manuscripts, samples, and a style file that generates the particular design specifications of that publication series. Though $\mathcal{AMS}$-LaTeX is the highly preferred format of TeX, author packages are also available in $\mathcal{AMS}$-TeX.

Authors may retrieve an author package for *Memoirs of the AMS* from www.ams.org/journals/memo/memoauthorpac.html or via FTP to ftp.ams.org (login as anonymous, enter username as password, and type cd pub/author-info). The *AMS Author Handbook* and the *Instruction Manual* are available in PDF format from the author package link. The author package can also be obtained free of charge by sending email to tech-support@ams.org (Internet) or from the Publication Division, American Mathematical Society, 201 Charles St., Providence, RI 02904-2294, USA. When requesting an author package, please specify $\mathcal{AMS}$-LaTeX or $\mathcal{AMS}$-TeX and the publication in which your paper will appear. Please be sure to include your complete mailing address.

After acceptance. The final version of the electronic file should be sent to the Providence office (this includes any TeX source file, any graphics files, and the DVI or PostScript file) immediately after the paper has been accepted for publication.

Before sending the source file, be sure you have proofread your paper carefully. The files you send must be the EXACT files used to generate the proof copy that was accepted for publication. For all publications, authors are required to send a printed copy of their paper, which exactly matches the copy approved for publication, along with any graphics that will appear in the paper.

Accepted electronically prepared files can be submitted via the web at www.ams.org/submit-book-journal/, sent via FTP, or sent on CD-Rom or diskette to the Electronic Prepress Department, American Mathematical Society, 201 Charles Street, Providence, RI 02904-2294 USA. TeX source files, DVI files, and PostScript files can be transferred over the Internet by FTP to the Internet node ftp.ams.org (130.44.1.100). When sending a manuscript electronically via CD-Rom or diskette, please be sure to include a message identifying the paper as a Memoir.

Electronically prepared manuscripts can also be sent via email to pub-submit@ams.org (Internet). In order to send files via email, they must be encoded properly. (DVI files are binary and PostScript files tend to be very large.)

Electronic graphics. Comprehensive instructions on preparing graphics are available at www.ams.org/authors/journals.html. A few of the major requirements are given here.

Submit files for graphics as EPS (Encapsulated PostScript) files. This includes graphics originated via a graphics application as well as scanned photographs or other computer-generated images. If this is not possible, TIFF files are acceptable as long as they can be opened in Adobe Photoshop or Illustrator. No matter what method was used to produce the graphic, it is necessary to provide a paper copy to the AMS.

Authors using graphics packages for the creation of electronic art should also avoid the use of any lines thinner than 0.5 points in width. Many graphics packages allow the user to specify a "hairline" for a very thin line. Hairlines often look acceptable when proofed on a typical laser printer. However, when produced on a high-resolution laser imagesetter, hairlines become nearly invisible and will be lost entirely in the final printing process.

Screens should be set to values between 15% and 85%. Screens which fall outside of this range are too light or too dark to print correctly. Variations of screens within a graphic should be no less than 10%.

Inquiries. Any inquiries concerning a paper that has been accepted for publication should be sent to memo-query@ams.org or directly to the Electronic Prepress Department, American Mathematical Society, 201 Charles St., Providence, RI 02904-2294 USA.

Editors

This journal is designed particularly for long research papers, normally at least 80 pages in length, and groups of cognate papers in pure and applied mathematics. Papers intended for publication in the *Memoirs* should be addressed to one of the following editors. The AMS uses Centralized Manuscript Processing for initial submissions to AMS journals. Authors should follow instructions listed on the Initial Submission page found at www.ams.org/memo/memosubmit.html.

Algebra to ALEXANDER KLESHCHEV, Department of Mathematics, University of Oregon, Eugene, OR 97403-1222; email: ams@noether.uoregon.edu

Algebraic geometry to DAN ABRAMOVICH, Department of Mathematics, Brown University, Box 1917, Providence, RI 02912; email: amsedit@math.brown.edu

Algebraic geometry and its applications to MINA TEICHER, Emmy Noether Research Institute for Mathematics, Bar-Ilan University, Ramat-Gan 52900, Israel; email: teicher@macs.biu.ac.il

Algebraic topology to ALEJANDRO ADEM, Department of Mathematics, University of British Columbia, Room 121, 1984 Mathematics Road, Vancouver, British Columbia, Canada V6T 1Z2; email: adem@math.ubc.ca

Combinatorics to JOHN R. STEMBRIDGE, Department of Mathematics, University of Michigan, Ann Arbor, Michigan 48109-1109; email: JRS@umich.edu

Commutative and homological algebra to LUCHEZAR L. AVRAMOV, Department of Mathematics, University of Nebraska, Lincoln, NE 68588-0130; email: avramov@math.unl.edu

Complex analysis and harmonic analysis to ALEXANDER NAGEL, Department of Mathematics, University of Wisconsin, 480 Lincoln Drive, Madison, WI 53706-1313; email: nagel@math.wisc.edu

Differential geometry and global analysis to CHRIS WOODWARD, Department of Mathematics, Rutgers University, 110 Frelinghuysen Road, Piscataway, NJ 08854; email: ctw@math.rutgers.edu

Dynamical systems and ergodic theory and complex analysis to YUNPING JIANG, Department of Mathematics, CUNY Queens College and Graduate Center, 65-30 Kissena Blvd., Flushing, NY 11367; email: Yunping.Jiang@qc.cuny.edu

Functional analysis and operator algebras to DIMITRI SHLYAKHTENKO, Department of Mathematics, University of California, Los Angeles, CA 90095; email: shlyakht@math.ucla.edu

Geometric analysis to WILLIAM P. MINICOZZI II, Department of Mathematics, Johns Hopkins University, 3400 N. Charles St., Baltimore, MD 21218; email: trans@math.jhu.edu

Geometric topology to MARK FEIGHN, Math Department, Rutgers University, Newark, NJ 07102; email: feighn@andromeda.rutgers.edu

Harmonic analysis, representation theory, and Lie theory to ROBERT J. STANTON, Department of Mathematics, The Ohio State University, 231 West 18th Avenue, Columbus, OH 43210-1174; email: stanton@math.ohio-state.edu

Logic to STEFFEN LEMPP, Department of Mathematics, University of Wisconsin, 480 Lincoln Drive, Madison, Wisconsin 53706-1388; email: lempp@math.wisc.edu

Number theory to JONATHAN ROGAWSKI, Department of Mathematics, University of California, Los Angeles, CA 90095; email: jonr@math.ucla.edu

Number theory to SHANKAR SEN, Department of Mathematics, 505 Malott Hall, Cornell University, Ithaca, NY 14853; email: ss70@cornell.edu

Partial differential equations to GUSTAVO PONCE, Department of Mathematics, South Hall, Room 6607, University of California, Santa Barbara, CA 93106; email: ponce@math.ucsb.edu

Partial differential equations and dynamical systems to PETER POLACIK, School of Mathematics, University of Minnesota, Minneapolis, MN 55455; email: polacik@math.umn.edu

Probability and statistics to RICHARD BASS, Department of Mathematics, University of Connecticut, Storrs, CT 06269-3009; email: bass@math.uconn.edu

Real analysis and partial differential equations to DANIEL TATARU, Department of Mathematics, University of California, Berkeley, Berkeley, CA 94720; email: tataru@math.berkeley.edu

All other communications to the editors should be addressed to the Managing Editor, ROBERT GURALNICK, Department of Mathematics, University of Southern California, Los Angeles, CA 90089-1113; email: guralnic@math.usc.edu.

Titles in This Series

941 **Gelu Popescu,** Unitary invariants in multivariable operator theory, 2009

940 **Gérard Iooss and Pavel I. Plotnikov,** Small divisor problem in the theory of three-dimensional water gravity waves, 2009

939 **I. D. Suprunenko,** The minimal polynomials of unipotent elements in irreducible representations of the classical groups in odd characteristic, 2009

938 **Antonino Morassi and Edi Rosset,** Uniqueness and stability in determining a rigid inclusion in an elastic body, 2009

937 **Skip Garibaldi,** Cohomological invariants: Exceptional groups and spin groups, 2009

936 **André Martinez and Vania Sordoni,** Twisted pseudodifferential calculus and application to the quantum evolution of molecules, 2009

935 **Mihai Ciucu,** The scaling limit of the correlation of holes on the triangular lattice with periodic boundary conditions, 2009

934 **Arjen Doelman, Björn Sandstede, Arnd Scheel, and Guido Schneider,** The dynamics of modulated wave trains, 2009

933 **Luchezar Stoyanov,** Scattering resonances for several small convex bodies and the Lax-Phillips conjuecture, 2009

932 **Jun Kigami,** Volume doubling measures and heat kernel estimates of self-similar sets, 2009

931 **Robert C. Dalang and Marta Sanz-Solé,** Hölder-Sobolv regularity of the solution to the stochastic wave equation in dimension three, 2009

930 **Volkmar Liebscher,** Random sets and invariants for (type II) continuous tensor product systems of Hilbert spaces, 2009

929 **Richard F. Bass, Xia Chen, and Jay Rosen,** Moderate deviations for the range of planar random walks, 2009

928 **Ulrich Bunke,** Index theory, eta forms, and Deligne cohomology, 2009

927 **N. Chernov and D. Dolgopyat,** Brownian Brownian motion-I, 2009

926 **Riccardo Benedetti and Francesco Bonsante,** Canonical wick rotations in 3-dimensional gravity, 2009

925 **Sergey Zelik and Alexander Mielke,** Multi-pulse evolution and space-time chaos in dissipative systems, 2009

924 **Pierre-Emmanuel Caprace,** "Abstract" homomorphisms of split Kac-Moody groups, 2009

923 **Michael Jöllenbeck and Volkmar Welker,** Minimal resolutions via algebraic discrete Morse theory, 2009

922 **Ph. Barbe and W. P. McCormick,** Asymptotic expansions for infinite weighted convolutions of heavy tail distributions and applications, 2009

921 **Thomas Lehmkuhl,** Compactification of the Drinfeld modular surfaces, 2009

920 **Georgia Benkart, Thomas Gregory, and Alexander Premet,** The recognition theorem for graded Lie algebras in prime characteristic, 2009

919 **Roelof W. Bruggeman and Roberto J. Miatello,** Sum formula for SL_2 over a totally real number field, 2009

918 **Jonathan Brundan and Alexander Kleshchev,** Representations of shifted Yangians and finite W-algebras, 2008

917 **Salah-Eldin A. Mohammed, Tusheng Zhang, and Huaizhong Zhao,** The stable manifold theorem for semilinear stochastic evolution equations and stochastic partial differential equations, 2008

916 **Yoshikata Kida,** The mapping class group from the viewpoint of measure equivalence theory, 2008

TITLES IN THIS SERIES

915 **Sergiu Aizicovici, Nikolaos S. Papageorgiou, and Vasile Staicu,** Degree theory for operators of monotone type and nonlinear elliptic equations with inequality constraints, 2008

914 **E. Shargorodsky and J. F. Toland,** Bernoulli free-boundary problems, 2008

913 **Ethan Akin, Joseph Auslander, and Eli Glasner,** The topological dynamics of Ellis actions, 2008

912 **Igor Chueshov and Irena Lasiecka,** Long-time behavior of second order evolution equations with nonlinear damping, 2008

911 **John Locker,** Eigenvalues and completeness for regular and simply irregular two-point differential operators, 2008

910 **Joel Friedman,** A proof of Alon's second eigenvalue conjecture and related problems, 2008

909 **Cameron McA. Gordon and Ying-Qing Wu,** Toroidal Dehn fillings on hyperbolic 3-manifolds, 2008

908 **J.-L. Waldspurger,** L'endoscopie tordue n'est pas si tordue, 2008

907 **Yuanhua Wang and Fei Xu,** Spinor genera in characteristic 2, 2008

906 **Raphaël S. Ponge,** Heisenberg calculus and spectral theory of hypoelliptic operators on Heisenberg manifolds, 2008

905 **Dominic Verity,** Complicial sets characterising the simplicial nerves of strict ω-categories, 2008

904 **William M. Goldman and Eugene Z. Xia,** Rank one Higgs bundles and representations of fundamental groups of Riemann surfaces, 2008

903 **Gail Letzter,** Invariant differential operators for quantum symmetric spaces, 2008

902 **Bertrand Toën and Gabriele Vezzosi,** Homotopical algebraic geometry II: Geometric stacks and applications, 2008

901 **Ron Donagi and Tony Pantev (with an appendix by Dmitry Arinkin),** Torus fibrations, gerbes, and duality, 2008

900 **Wolfgang Bertram,** Differential geometry, Lie groups and symmetric spaces over general base fields and rings, 2008

899 **Piotr Hajłasz, Tadeusz Iwaniec, Jan Malý, and Jani Onninen,** Weakly differentiable mappings between manifolds, 2008

898 **John Rognes,** Galois extensions of structured ring spectra/Stably dualizable groups, 2008

897 **Michael I. Ganzburg,** Limit theorems of polynomial approximation with exponential weights, 2008

896 **Michael Kapovich, Bernhard Leeb, and John J. Millson,** The generalized triangle inequalities in symmetric spaces and buildings with applications to algebra, 2008

895 **Steffen Roch,** Finite sections of band-dominated operators, 2008

894 **Martin Dindoš,** Hardy spaces and potential theory on C^1 domains in Riemannian manifolds, 2008

893 **Tadeusz Iwaniec and Gaven Martin,** The Beltrami Equation, 2008

892 **Jim Agler, John Harland, and Benjamin J. Raphael,** Classical function theory, operator dilation theory, and machine computation on multiply-connected domains, 2008

891 **John H. Hubbard and Peter Papadopol,** Newton's method applied to two quadratic equations in $\mathbb{C}^2$ viewed as a global dynamical system, 2008

890 **Steven Dale Cutkosky,** Toroidalization of dominant morphisms of 3-folds, 2007

889 **Michael Sever,** Distribution solutions of nonlinear systems of conservation laws, 2007

For a complete list of titles in this series, visit the
AMS Bookstore at **www.ams.org/bookstore/**.